Optical Applications of Liquid Crystals

T0203912

Series in Optics and Optoelectronics

Series Editors: **R G W Brown**, University of Nottingham, UK
E R Pike, Kings College, London, UK

Series in Optics and Optoelectronics

Optical Applications of Liquid Crystals

Edited by

L Vicari
Università di Napoli 'Federico II', Napoli, Italy
INFM – Istituto Nazionale per la Fisica della Materia

CRC Press
Taylor & Francis Group
Boca Raton London New York

CRC Press is an imprint of the
Taylor & Francis Group, an **informa** business

First published 2003 by IOP Publishing Ltd

Published 2019 by CRC Press
Taylor & Francis Group
6000 Broken Sound Parkway NW, Suite 300
Boca Raton, FL 33487-2742

First issued in paperback 2019

ISBN-13: 978-0-367-45449-4 (pbk)
ISBN-13: 978-0-7503-0857-1 (hbk)

Visit the Taylor & Francis Web site at
http://www.taylorandfrancis.com

and the CRC Press Web site at
http://www.crcpress.com

British Library Cataloguing-in-Publication Data

A catalogue record for this book is available from the British Library.

Library of Congress Cataloging-in-Publication Data are available

Cover Design: Victoria Le Billon

Typeset by Academic + Technical Typesetting, Bristol

Contents

4 Polymer-dispersed liquid crystals   148
F Bloisi and L Vicari
Università 'Federico II' di Napoli, Italy
INFM – Istituto Nazionale per la Fisica della Materia

Preface

Liquid crystals (LCs) are materials showing characteristics that are inter-mediate between those of a crystal and those of an isotropic liquid, thus possessing unique electric and optical properties. Even though LCs have been known for more than a century, they have become widely used in electro-optic applications only since the early 1960s. In the past few decades there has been growing interest in the field: it has become a hot research topic in physics and chemistry, and many technical papers, several conference proceedings and many books have been written concerning LCs and their applications, a research field now involving hundreds of physicists, chemists and engineers. The reader interested in LCs can find several excellent books concerning the physical and/or chemical basis of the field and some general handbooks. However, if he is interested in the applications he has to look mainly at technical papers or conference proceedings. This book attempts to fill this gap, presenting current and future applications from the point of view of the main researchers coming from both industry and universities. Part of this book is a collection of chapters written by researchers coming from the leading industries operating in the field. They present the main current applications of LCs (three-dimensional holographic displays, projec-tion displays, telecommunication devices, optical data processing devices, adaptive optics devices, spatial light modulators, etc.). The remaining chapters, written by well-known university researchers, are devoted to discussing the more promising technologies (photo aligning, photo pattern-ing, polymer dispersed liquid crystals, etc.). The book is devoted to the most recent developments in the use of liquid crystals for optical application in industrial and scientific instrument devices. The arguments of each chapter are treated at a specialist level; however, each chapter is organized in such a way as to be reader friendly for a non-specialized reader. Besides an intro-duction, introducing any advanced physical concept used in the chapter, and a bibliography including both current technical papers and general books, each chapter uses clear figures and good quality photos with detailed captions to be accessible to readers coming from a non-specialized field of interest and to graduate students. Some examples or case studies help one to understand problem complexity, while tables, reporting the physical and chemical properties, help to develop self-organized examples.

L Vicari

Chapter 1

Optical properties and applications of ferroelectric and antiferroelectric liquid crystals

Emmanouil E Kriezis, Lesley A Parry-Jones
and Steve J Elston

1.1 Introduction

1.1.1 Smectic liquid crystals

The most simple liquid crystal phase is the nematic phase, in which the molecules possess orientational ordering (like a crystal) but no positional ordering (like a liquid). However, there are some liquid crystal phases in which the molecules do exhibit a degree of positional ordering. In the smectic phases, this positional ordering is in one dimension only, forming layers of two-dimensional nematic liquids.

The most simple smectic phase is the smectic A (SmA) phase, in which the direction of average molecular orientation (the director, **n**) is along the smectic layer normal (see figure 1.1(a)). In addition, there is a family of tilted smectic liquid crystal phases, in which the director is at a fixed angle θ with respect to the layer normal. Each phase differs in the relationship between the azimuthal angles of the director in adjacent layers. The simplest cases are the smectic C and smectic C_A phases (SmC and SmC_A), which are illustrated in figures 1.1(b) and (c), respectively. In the SmC phase, the director is constant from one layer to the next, whereas in the SmC_A phase the director alternates in tilt direction from layer to layer, forming a herring-bone structure.

When the phases comprise chiral molecules, chiral versions of these phases are formed: SmA^*, SmC^* and SmC_A^*. One of the effects of the chirality of the molecules in the case of the tilted smectics SmC^* and SmC_A^* is to cause the azimuthal angles of the directors to precess slowly from one layer to the next. This creates a macroscopic helical structure

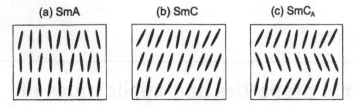

Figure 1.1. The three most simple smectic phases (a) SmA, where the director is perpendicular to the layer normal, (b) SmC where the director tilts at a constant angle θ to the layer normal, and (c) SmC_A where the direction of tilt alternates from one layer to the next, forming a herringbone structure.

with its axis along the layer normal, which tends to have a pitch of around 100–1000 layers.

1.1.2 Typical molecular structure

Like a typical nematic liquid crystal molecule, a smectic mesogen comprises a rigid core (which tends to be made up of two or three ring structures) with flexible chains at either end. For example, 8CB (see figure 1.2(a)) forms a SmA phase below its nematic phase. DOBAMBC, a typical SmC^* material, and MHPOBC, a typical SmC_A^* material, are illustrated in figures 1.2(b) and (c), respectively. In order to form a layered, smectic phase, there must be significant intermolecular interactions (either hydrogen or van der Waals in origin) [1].

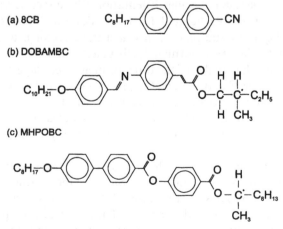

Figure 1.2. Typical molecular structure of smectic liquid crystals, as exemplified by (a) 8CB, which forms an SmA phase, (b) DOBAMBC, which forms an SmC^* phase, and (c) MHPOBC, which forms an SmC_A^* phase.

1.1.3 Order parameters

Just like the orientational ordering in a nematic liquid crystal, the positional ordering of a smectic material is not perfect. In some cases a plot of the density of the molecular centres of mass as a function of distance along the normal to the layers, x, follows a sinusoidal variation,

$$\rho(x) = \rho_0\left(1 + \psi \sin\left(\frac{2\pi x}{\delta}\right)\right),$$

where ρ_0 is the mean density and δ is the layer spacing, which is typically a few nanometers. ψ is the smectic order parameter, which is the ratio of the amplitude of oscillation to the mean layer density, and hence expresses the extent to which the material is layered, typically $\psi \ll 1$.

Within each layer, the orientational ordering about the director is characterized by a nematic-like order parameter, S, defined as

$$S = \langle P_2(\cos\omega)\rangle = \langle \tfrac{3}{2}\cos^2\omega - \tfrac{1}{2}\rangle,$$

where ω is the angle between the molecule and the director, $\mathbf{n}$, and P_2 is the second-order Legendre polynomial. S is an ergodic variable, so that the average can be performed either over many molecules at one point in time, or for one molecule over a period of time.

The order parameters S and ψ are sufficient to describe the SmA phase. However, for the tilted smectic phases, two further order parameters are required to fully describe the phase: the tilt of the director with respect to the layer normal (usually given the symbol θ), and the azimuthal angle of the director with respect to some fixed coordinate system (often given the symbol ϕ). These angles are illustrated in figure 1.3.

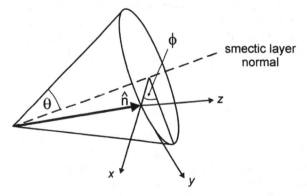

Figure 1.3. The smectic cone: an illustration of the tilt (θ) and azimuthal (ϕ) angles in a tilted smectic liquid crystal layer. θ is the half angle of the smectic cone, and ϕ describes the position of the director on the surface of the cone, with respect to some (arbitrary) reference point. Also shown are a set of coordinate axes, x, y and z, which are useful for defining the physical properties of the material.

1.1.4 Point symmetries of the smectic phases

As well as translational symmetry, the SmA phase has the following point symmetries:

1. mirror symmetry about any plane parallel to the smectic layers that is either exactly between planes or exactly midplane;
2. two-fold rotational symmetry about any axis contained within any of the above mirror planes;
3. mirror symmetry about any plane perpendicular to the smectic layers;
4. complete rotational symmetry about the axis perpendicular to the layers.

This set of point symmetries corresponds to the symmetry $D_{\infty h}$ in the Schoenflies notation. The chiral version of the SmA phase (SmA*) has only the rotational symmetries in the list above: the mirror symmetries no longer exist because the constituent molecules are chiral. This reduces the symmetry of the SmA* phase to D_{∞}.

The high symmetry of the SmA and SmA* phases precludes the existence of any net spontaneous polarization, just like a nematic. They can therefore only respond to an applied field via an induced electric dipole.

The point symmetries of the SmC phase are as follows:

1. mirror symmetry in the tilt plane of the molecules;
2. two-fold rotational symmetry about the axis perpendicular to the tilt plane of the molecules, either exactly between layers or exactly mid-layer.

This combination of point symmetries corresponds to the C_{2h} symmetry group in the Schoenflies notation, and also preclude the existence of any net spontaneous polarization in the SmC liquid crystal phase.

However, in the chiral version of the SmC phase (SmC*), the mirror symmetry is no longer present (due to the chirality of the molecules) and hence only the rotational symmetry remains. The symmetry group is reduced to C_2, and it is hence possible for a spontaneous polarization to exist along the C_2 axis, that is, along the direction of the two-fold rotation axis.

1.1.5 Ferroelectricity and antiferroelectricity in liquid crystals

As discussed above, the symmetry of the SmC* phase is such that a spontaneous polarization is permitted along the C_2 axis of each smectic layer. The net spontaneous polarization arises due to the lack of rotational degeneracy of the molecules about their long axes within the smectic layer. To understand this, it is first constructive to consider the SmA* phase. In general, each molecule has an electric dipole along an arbitrary direction. The symmetry is such that there are as many molecules pointing 'up' within a layer as there are pointing 'down' within a layer, and hence there can be no net component of the polarization perpendicular to the layers. There is also complete rotational symmetry about an axis perpendicular to the layers, which means that there can be no net polarization component within the smectic layer. Hence the SmA^* phase has no net electric dipole.

However, if the molecules now tilt with respect to the layer normal (forming the SmC^* phase), the symmetry of the system changes. Whatever the mechanism for causing the molecules to tilt with respect to the layer normal, there must be a break in the rotational degeneracy of the molecules about their axes in order to cause that change. It is this hindered rotation of the molecules about their long molecular axis that allows a net spontaneous polarization to exist perpendicular to the molecular tilt plane. Hence, as predicted by Meyer *et al* [2], SmC^* liquid crystals are ferroelectric.

In fact, this is not strictly true, as only a single smectic layer has been considered here. As noted above, the chirality of the molecules also causes a macroscopic helical structure to exist, such that the C_2 axis (and hence the polarization direction) precesses slowly from one layer to the next. Thus on a macroscopic level there is no net polarization, and a more correct name for the phase is therefore 'helielectric'. However, in many instances of the use of SmC^* liquid crystals, they are confined to a cell geometry in such a way that the helical structure is suppressed (known a surface stabilization, see section 1.3.6), and then the system is truly ferroelectric.

Following the same chain of argument, SmC_A^* materials are anti-ferroelectric. This is because each smectic layer, considered individually, has the same symmetry as a SmC^* layer. Therefore each layer has a net spontaneous polarization along its C_2 axis. However, because of the alternating tilt directions in adjacent layers, the polarizations also alternate in direction from layer to layer, and hence the material is described as being antiferroelectric.

1.2 Material properties

1.2.1 Optical properties

The orientational order of liquid crystals combined with their molecular anisotropy leads to anisotropic physical properties (including the optical properties), just like those in an anisotropic crystalline solid.

The refractive index of a nematic liquid crystal is different along the average molecular direction (the director) compared with that along all directions perpendicular to the director. The nematic phase, for example, has uniaxial symmetry because the optical indicatrix has only one isotropic plane. The system is uniaxial because of the complete rotational degeneracy about the director. The same is true of the SmA phase. In the SmC phase, however, the index ellipsoid does not have complete rotational symmetry about any axis because the phase is biaxial. This is because of the hindered molecular rotation about the long axis. Thus there are three distinct refractive indices.

Note that the symmetry of the phase is not necessarily the same as that of the constituent molecules. For example, a uniaxial phase can be formed with biaxial molecules (e.g. the SmA phase), and a biaxial phase can be formed with uniaxial molecules (e.g. the SmC_A phase).

Any chiral liquid crystal phase will also exhibit optical activity. The rotation of linearly polarized light comes from two different sources. The first is inherent to the chirality of the molecules and is present even in the isotropic phase (e.g. a sugar solution is optically active). The second is caused by the macroscopic helical structure that tends to be formed by chiral mesogens. Depending on the refractive indices, the pitch of the helix and the wavelength of light used, it is possible for the eigenmodes of the system to be right and left circularly polarized light, and therefore for the system to strongly rotate the plane of linearly polarized light.

1.2.2   Dielectric properties

In the same way that the optical properties of liquid crystals are anisotropic, so too are the dielectric properties: the polarization induced by an applied electric field depends on the direction of the applied field. This anisotropy is expressed in the dielectric permittivity tensor, $\tilde{\varepsilon}$:

$$\mathbf{D} = \tilde{\varepsilon}\mathbf{E}; \qquad \begin{pmatrix} D_x \\ D_y \\ D_z \end{pmatrix} = \begin{pmatrix} \varepsilon_{xx} & \varepsilon_{xy} & \varepsilon_{xz} \\ \varepsilon_{yx} & \varepsilon_{yy} & \varepsilon_{yz} \\ \varepsilon_{zx} & \varepsilon_{zy} & \varepsilon_{zz} \end{pmatrix} \begin{pmatrix} E_x \\ E_y \\ E_z \end{pmatrix}.$$

Symmetry considerations lead to the fact that the permittivity tensor must be symmetric, i.e. $\varepsilon_{ij} = \varepsilon_{ji}$ [3]. If the axes of the coordinate system are aligned with the eigenvectors of the liquid crystal system, then the off-diagonal terms are zero, and only ε_{xx}, ε_{yy} and ε_{zz} need be considered.

For a uniaxial phase, such as the nematic phase, with the director along the z axis, $\varepsilon_{xx} = \varepsilon_{yy}$. The dielectric anisotropy in this case is defined to be

$$\Delta\varepsilon = \varepsilon_{zz} - \varepsilon_{xx}.$$

In the biaxial SmC phase, ε_{xx}, ε_{yy} and ε_{zz} are all different, and hence it is not possible to identify a single dielectric anisotropy. However, referring to figure 1.3, which defines the conventional coordinate axes, it is possible to define a 'uniaxial anisotropy',

$$\Delta\varepsilon = \varepsilon_{zz} - \varepsilon_{xx},$$

and a dielectric biaxiality,

$$\delta\varepsilon = \varepsilon_{yy} - \varepsilon_{xx}.$$

These conventions [4] assume that the frequency of the applied electric field is low (essentially static) so that none of the dielectric modes are relaxed out. At optical frequencies ($\sim 10^{15}$ Hz) the dielectric constants are much smaller (due to relaxation) and are related to the refractive indices of the material:

$$n_e = \sqrt{\varepsilon_{zz}}, \qquad n_o = \sqrt{\varepsilon_{xx}}, \qquad \Delta n = \sqrt{\varepsilon_{zz}} - \sqrt{\varepsilon_{xx}},$$

for a uniaxial system. Within the optical regime, the refractive indices and dielectric permittivities are frequency dependent, i.e. dispersive. It is generally the case that the parameters decrease with wavelength, as does the birefringence.

1.2.3 Mechanical properties

The optical and dielectric properties of liquid crystals, discussed above, are generally determined by the rigid core of the molecules. This is because the core generally contains two or more benzene rings which have a high density of delocalized electrons. The core is therefore highly polarizable and the dominant part of the molecule in terms of optical and dielectric properties. However, the mechanical properties of the liquid crystal are determined by the entire length of the molecule. The tilt angles θ observed using optical and mechanical (x-ray diffraction) methods are sometimes very different. This has often been attributed to the difference in tilt angle between the core and whole length of the molecule [5, 6] (see figure 1.4). This effect is important, since it explains the anomalous crossover in optical and mechanical tilt angles with temperature in some materials, in terms of a conformational change in the molecular structure (see figure 1.4). However,

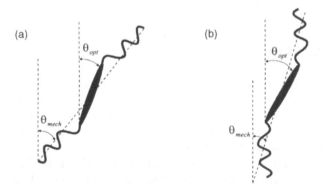

Figure 1.4. Illustration of the difference between the optical and mechanical tilt angles that can exist in tilted smectic liquid crystals, if the molecules have a zig-zag shape. Sometimes the shape of the molecule is temperature-dependent, and conformational changes can occur, so that, for example (a) at high temperatures the optical tilt angle is less than the mechanical tilt angle, and (b) at lower temperatures the opposite is the case.

for most materials the optical tilt angle is generally greater than the mechanical tilt angle, and this can be explained by alternative theories.

The mechanical tilt angle is determined by measuring the smectic layer spacing (via x-ray powder diffraction) in the SmA and SmC phases, and using the following formula:

$$\theta_{\text{mechanical}} = \cos^{-1}(d_a/d_c).$$

However, this method assumes that the aspect ratio of the molecules is infinite, and that they are perfectly ordered about the director ($S = 1$). Under these assumptions, the layer contraction in going from the SmA to the SmC phase would be given by the cosine of the tilt angle, and the equation above would be correct. However, due to the finite molecular width and imperfect molecular ordering, the true tilt angle tends to be somewhat larger than the layer contraction would indicate. In order to understand why the molecular width is important, consider the extreme case in which the molecules are spherical. Then there would be no layer contraction as the molecules tilt with respect to the layer normal. For more realistic molecules, which have a somewhat higher aspect ratio, we can expect the layer contraction angle to be about 90% of the true tilt angle [7]. The degree of molecular ordering is also important, as can be understood by considering the de Vries [8] model of the SmA phase in which the molecules are all tilted at a fixed angle with respect to the layer normal, but there is complete azimuthal degeneracy so that the director is still along the layer normal. In transforming to the SmC phase, all that happens is that the azimuthal distribution of the molecules biases towards one side of the layer normal. In this extreme case there would be no change in the layer thickness. This effect may account for some of the discrepancy observed in the observed mechanical and optical tilt angles.

1.3 Alignment

Of key importance to the use of any liquid crystal material in an electro-optic device is the issue of alignment [9]. Not all electro-optic applications require a highly aligned liquid crystal state; for example, some systems may require the material to be initially in a randomly aligned state in order to strongly scatter light. However, the majority of important applications are based on the interaction of the light with the optical anisotropy of the liquid crystal in a controlled way, and therefore a well aligned initial state is desirable. While this is relatively easy to achieve in a liquid crystal device filled with material in the nematic phase, it is somewhat more difficult to construct devices containing smectic liquid crystals which are well aligned. This is due to the competing issues of alignment of the molecular axis and alignment of the smectic layers. If these issues are not resolved through the formation of a well defined layer structure then it is common for defects to form, which can significantly influence the electro-optic properties of the device.

1.3.1 Homogeneous, homeotropic and pretilt alignment

The two most common alignments which can form in a liquid crystal device are referred to as homogeneous and homeotropic alignment (figure 1.5). In homogeneous alignment the internal surfaces of a liquid crystal containing device are treated to cause the molecular axis to lie parallel, or nearly parallel, to the surface. For homeotropic alignment, however, the surfaces are treated to cause the molecular axis to lie perpendicular, or nearly perpendicular, to the surface. While both of these alignment treatments are interesting from the point of view of their physics it is generally homogeneous alignment which leads to interesting electro-optic properties which can be exploited in device technology.

Homogeneous alignment can be achieved in a number of ways, the most common of which is to deposit a thin layer of polymeric material on the surface and then to physically rub this to create an ordered surface. This has resulted in the alignment direction being commonly referred to as the rubbing direction. The interaction between the ordered surface which is formed and the liquid crystal in contact with it causes the liquid crystal to be aligned. This alignment does not necessarily rely on the physical scouring of the polymer surface but depends on the interaction between the liquid crystal molecules and the aligned polymer molecules on the surface. A similar effect has been achieved by the curing of a polymer surface under polarized ultraviolet light [10]. In most cases the liquid crystal alignment achieved is not perfectly planar, i.e. the liquid crystal director does not lie perfectly

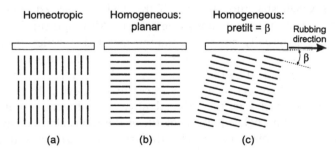

Figure 1.5. Alignment in a liquid crystal, showing (a) homeotropic alignment, where the molecular axis is perpendicular to the substrate surface; (b) homogeneous planar alignment, where the molecular axis is parallel to the substrate surface; (c) homogeneous pretilted alignment, where the molecular axis is tilted by a few degrees, β, from the substrate surface. The structure illustrated is in effect that of a layered smectic A liquid crystal material.

parallel to the surface. Normally it is tilted away from the surface by a few degrees, with typical pretilts being in the range of 2° to 4° (figure 1.5(c)).

1.3.2 Ideal bookshelf

A basic smectic material typically has a phase sequence (on cooling) of isotropic, nematic, SmA, SmC. In the nematic phase we would expect such a material to form a simple uniformly aligned layer if placed in a device with parallel-aligned homogeneous surfaces. Such a layer will behave as a simple wave plate, with its order depending on the thickness of the nematic layer. If the material is strongly chiral then a twisted structure can be formed, but provided the full helical pitch is at least four times the device thickness then a uniform layer should form as described.

When such a structure is further cooled into the SmA phase the uniform alignment is normally retained. In fact, due to the rather rigid nature of the smectic layering which forms, the alignment commonly improves in the SmA phase. This is because small defects in the surface alignment, which propagate into the bulk of the nematic structure, do not significantly influence the SmA structure due to the presence of the smectic layers. In this state the nematic-like director remains aligned with the direction originally imposed by the surfaces and the smectic layers form perpendicular to this direction.

If a uniform SmA structure is cooled into the SmC phase then two competing influences of the smectic layer structure and the director-surface interaction are present. In the simplest case the smectic layer structure may remain the same as that present in the SmA phase, and then the director structure must take a form consistent with this (see figure 1.6). In this case it is assumed that the smectic layers remain perpendicular to the surfaces and the original alignment direction. This structure is commonly referred to as a bookshelf structure, because if the device is viewed from the side the smectic layers appear as books on a shelf. Although the formation of this structure is actually not very common in tilted smectics, as explained below, it is very often assumed to be present. For this structure there are two stable states defined by the intersection of the smectic cone (the allowed positions of the director) with the surfaces of the device. In this case the director no longer lies in the original alignment direction, but is tilted away from it by an angle equal to the tilt of the molecular axis away from the smectic layer normal. Such devices are further discussed in section 1.5.2.1.

1.3.3 Chevron formation in tilted smectics

Although the smectic layer structure described above in section 1.3.2 occurs naturally in some tilted smectic materials, and can be accessed in others through the application of large electric fields, it is not the most commonly

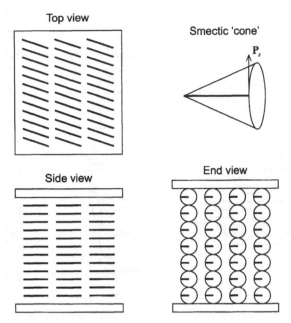

Figure 1.6. The so-called 'bookshelf structure' in a ferroelectric liquid crystal. The smectic layers are perpendicular to the device surfaces and the molecular axis tilts in the plane of the surfaces.

occurring structure. This is because as materials cross from the SmA into the SmC phase during cooling the director tilt is commonly accompanied by shrinkage of the smectic layer thickness, as described in section 1.2.3.

Assuming that in the SmA phase the smectic layers have formed perpendicular to the device surfaces, then when crossing into the SmC phase there are two possibilities: (i) the smectic layers could rearrange in a way consistent with retention of their bookshelf-like structure, as described above, although there is no obvious force driving this; (ii) the smectic layers could rearrange by retaining their periodicity in the layer structure formed in the device in the SmA phase—importantly this does not require flow of material between the smectic layers.

In practice it is generally the latter which occurs through a tilting of the smectic layering. In order to minimize flow of the material at the device surfaces, this tilting generally occurs in opposite directions in the top and bottom halves of the device, and a kink forms in the centre. This results in what has commonly been termed a 'chevron structure' in the smectic layering (figure 1.7). The outcome is that through the SmA to SmC phase transition there is no flow of material at the surfaces and no flow of material between smectic layers. As explained in section 1.2.3, the tilt angle of the smectic layers, δ, is typically 10% smaller than the tilt of the molecular axis away from the smectic layer normal, θ.

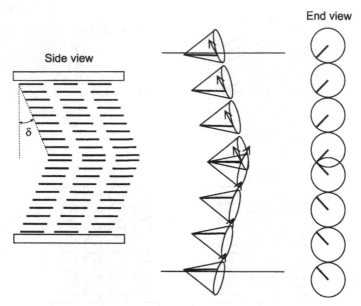

Figure 1.7. The basic chevron structure in ferroelectric liquid crystals. This occurs due to layer thinning across the smectic A to smectic C phase transition. There are a number of stable states of the director structure, the most simple of which is that shown (and the version of this tilted in the opposite direction). However, switching at the surfaces and/ or chevron interface can also lead to a number of complex half-splayed (or half-twisted) states.

Due to the small difference between the molecular tilt angle and the tilt in the smectic layering, the structure which forms at the kink (or chevron interface) in the centre of the device is quite complex. In general there are two possible places where the molecular axis can lie at this point. These are defined by the intersection of the smectic cones for the two halves of the chevron structure (see figure 1.7). Switching between these two states is a key part of the behaviour of devices with a chevron structure in the smectic layering.

1.3.4 Influence of pretilt

Having discussed how the chevron structure forms due to the changes in smectic layer thickness in the tilted smectic phases, it is also necessary to consider the surface interaction. As in the case of the bookshelf structure discussed above it is generally the intersection between the 'smectic cone' of allowed director orientations and the plane of the surface which dictates the surface states. Therefore if, as outlined above, the smectic layer tilt angle is a little less than the smectic cone angle (typically by a few degrees at room temperature) there are two intersections between the cone and the surface. These define two surface switched states. In principle it is then possible for a SmC material to have two possible states at each surface and two possible states at the chevron interface, leading to a number of configurations. This is, however, not entirely desirable, and it is generally possible to overcome this through the introduction of an appropriate surface pretilt.

For certain surface alignment treatments the molecular axis at the surface does not lie in the plane of the surface, but is tilted away from this by a few degrees. This is quite common in the rubbed polymer alignment treatments normally used to align liquid crystals. Now, if this pretilt is arranged to be equal to the difference between the smectic cone angle and the smectic layer tilt angle then a unique situation can occur. In this case the molecular axis at the surface, which is constrained to lie on the smectic cone, can also lie in the original alignment direction. In this case only a single surface state exists and all switching takes place at the chevron interface. Such a structure is referred to as the C2 structure, and in the relaxed state the director has a roughly triangular profile through the thickness of the device (figure 1.8(a)).

One can see that due to the surface pretilt the symmetry in the tilt direction of the smectic layers is broken. That is, the smectic layers have to tilt in a well-defined direction for the difference between the smectic cone angle and smectic layer tilt angle to match the surface pretilt angle. If the smectic layers tilt in the opposite direction then a matching condition does not occur and a different structure forms (see figure 1.8(b)). This is referred to as the C1 structure because in a typical device it is the structure which forms at higher temperature, whereas at lower temperatures the C2 structure forms.

1.3.5 Zig-zag formation

In a typical device it is possible for the chevron structure to form in two possible directions with respect to the rubbing direction, forming either a C1 or C2 structure. This results in interface regions between these tilt directions, which appear as defects within a device. These are commonly referred to as zig-zag defects due to their appearance when viewed under a polarizing microscope. Their formation is generally detrimental to the behaviour of a device for a number of reasons: (i) they tend to scatter light and therefore reduce optical contrast; (ii) the switching properties tend to be different on either side of the defect as one side is in a C1 state and the other in a C2 state; (iii) for small pixels the area of the zig-zag defect wall can be a significant proportion of the pixel's area, reducing the effective area of switching liquid crystal.

Generally the formation of zig-zag defects is avoided by appropriate choice of the surface pretilt angle [11]. Either the surface pretilt angle is chosen to match the difference between the smectic cone angle and the

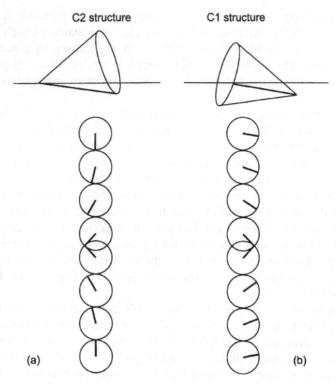

Figure 1.8. The surface states, and end views of the internal structure, for the C2 and C1 structures which can occur in a chevron layer formation in a device with a degree of surface pretilt. Depending on which way the smectic layers tilt with respect to the surface tilt the alignment at the surfaces can move either towards (C2) or away from (C1) the original surface alignment direction.

smectic layer tilt angle at room temperature in order to encourage the formation of a C2 structure, or the pretilt is chosen to be large enough to ensure that the device remains in a C1 state. As noted above, it is normally the C2 structure, which is preferred, due to the simplicity of having only one switching interface, which is at the kink in the chevron structure. If a C1 structure was chosen, there would be three switching interfaces, at the two surfaces as well as the chevron tip, resulting in a far more complex system.

1.3.6 Surface stabilization

In the structures described above there has been no mention of how the helical nature of chiral smectics influences the behaviour of devices. It has been assumed that in the device the material is not in a helical state. This is often true, but depends critically on the relation between the natural helical pitch of the material and the thickness of the device in which it is used.

Typically, if the device is thinner than approximately the natural helical pitch of the material, then the helical structure is suppressed. However, if the device is thicker than this, then the device tends to be in a helical state [12–14]. This leads to a complex internal director structure due to the interaction between the bulk helical state and the planar surface of the device.

1.4 Optical properties of smectic structures

1.4.1 Optics of simple wave-plate

If a simple (ideal) bookshelf structure is formed in the smectic layering, either through appropriate choice of material or by suitable post-processing treatment of a device, then the optical properties are generally quite simple. As noted above in section 1.3.2, there are two stable states defined by the intersection between the smectic cone and the surfaces of the device (provided that the device is thin enough to suppress the helical structure). A common configuration of such a device is to place it between crossed polarizers, in which case the transmission is given by

$$T = \sin^2(2\chi) \sin^2 \left(\frac{\pi \Delta n d}{\lambda} \right), \qquad (1.1)$$

where χ is the angle between the optic axis and either polarizer, Δn is the birefringence of the material, d is the device thickness and λ is the wavelength of light. In practice this is an approximation to the true transmission, due both to the influence of imperfect alignment on the liquid crystal structure and also to the effects of reflection within the device (such as at the glass–liquid crystal interface) which are not included in equation (1.1). It should also be noted that liquid crystal materials are highly dispersive, and therefore the variation of Δn is important. Across the visible spectrum (from 400 to 700 nm) Δn can typically vary from 0.18 to 0.14.

1.4.2 Optical properties of chevron structures without pretilt

When the smectic layering takes on a chevron structure then a number of director states are possible. With simple planar surfaces the simplest of these has a reasonably uniform director structure, where the intersections of the smectic cones with the surfaces and the intersection at the chevron interface are all approximately coplanar. However, the difference in optical tilt angle between the two states is considerably reduced from that in the bookshelf structure, and is typically between one half and one third of that available in the bookshelf structure.

   In a chevron structure it is quite common for the effects of surface polar interaction to lead to states at the two surfaces where the dipole preferentially points towards (or away from) the surface. This causes the director on each surface to lie on opposite sides of the smectic cone. In the case of a chevron structure the twisted structure then becomes a half-twisted (or half-splayed) structure because of the additional effect of the chevron interface. For the half of the device where the surface orientation is on the same side as the chevron intersection orientation the structure will be approximately uniform. However, for the other half of the device the chevron intersection orientation will be on the opposite side of the smectic cone from the surface orientation. Therefore in this half of the device a partially twisted (or splayed) structure is formed. The optics of these states is quite complex, and this means that the structures are not generally useful in device applications.

1.4.3 Optical properties of chevron structures with pretilt: C1 and C2 states

As outlined above (section 1.3.5), if the surface pretilt is chosen appropriately then a C2 chevron state can be achieved. In this case the director lies approximately in the plane of the cell surfaces and has an in-plane tilt angle which is roughly triangular in profile. At the surfaces the director is aligned in the natural alignment direction. As the device is traversed from one surface to the other it tilts out to the intersection point at the chevron interface and back to the alignment direction. The tilt direction can be to either side of the alignment direction, depending on which state the device has previously been switched to. In this case the apparent tilt angle depends strongly on the wavelength of light. For long wavelength light an average tilt effect is seen. However, for short wavelength light a Mauguin limit type guiding effect can take place in the polarization and the effective tilt angle can be zero; i.e. whichever state the device is switched to the apparent alignment direction is the same. For typical device thicknesses at visible wavelengths the apparent tilt angle is around one third of the smectic cone angle.

If some surface pretilt is present and a C1 structure is formed in the smectic layering, then the states are quite similar to those described for a chevron structure with planar surfaces. Commonly the surface polar interaction with the material's spontaneous polarization leads to half-twisted (or half-splayed) states. Again these have complex optical properties, which are not generally useful in device applications.

1.4.4 Optical properties of antiferroelectric liquid crystal structures

In a sufficiently thin antiferroelectric liquid crystal cell the helical structure is suppressed and the simplest structure which can then form is one where the molecular tilt plane lies parallel to the surface (see section 1.5.5.3). In this structure the molecular axes in alternate layers tilt in opposite directions from the layer normal, by an amount equal to the smectic cone angle. However, the director remains in the plane of the cell surfaces throughout.

The result is a structure which macroscopically is optically biaxial. This is because the thickness of the individual smectic layers is much less than the wavelength of light. Matching the field conditions at the layer interfaces allows the effective refractive indices both along the layer normal, n_{normal}, and perpendicular to that but in the plane of the cell surfaces, $n_{parallel}$, to be determined:

$$n_{normal} = \sqrt{n_e^2 \cos^2 \theta + n_o^2 \sin^2 \theta},$$

$$n_{parallel} = \frac{n_o n_e}{\sqrt{n_e^2 \cos^2 \theta + n_o^2 \sin^2 \theta}}.$$

The refractive index perpendicular to the cell surface is simply equal to the ordinary refractive index of the material, n_o.

If the device is not sufficiently thin to suppress the helix then the bulk structure of the antiferroelectric layer is of course helical. Due to the alternation of the molecular axis tilt angle from layer to layer the material remains locally biaxial. It therefore appears as a helical biaxial structure, with a pitch of one half of the helical pitch of the liquid crystal itself. This structure can lead to diffraction of light. How this structure responds to applied electric fields, and the consequences for its optical properties, are discussed in section 1.5.5.2.

1.5 Interaction with electric fields

The response of liquid crystals to applied electric fields is very important, as it is this that will determine the electro-optic properties of the material. Broadly speaking, the response is fundamentally the same in all liquid crystals: the director is reoriented in order to maximize the alignment of the polarization **P** with the applied field **E**, as this minimizes the electric energy density, $-\mathbf{P} \cdot \mathbf{E}$. However, the way in which the polarization arises (i.e. whether it is spontaneous or induced) and how the reorientation affects the liquid crystal structure can be quite different, as the following examples will illustrate. Unless otherwise stated, the liquid crystals are assumed to be in the bookshelf geometry, with the electric field applied parallel to the smectic layers.

1.5.1 SmA* field response

In SmA* materials which also have a SmC* phase, there is an interesting field response of the SmA* phase near the SmA*–SmC* phase transition, known as the electroclinic effect or soft mode [15].

In the SmA* phase there is no net spontaneous polarization because of the complete rotational degeneracy about the smectic layer normal. As the temperature is reduced, the amount of rotation about the long molecular axes decreases, and simultaneously, biaxiality, net polarization and a molecular tilt appear at the phase transition to SmC*. It is also possible to induce the SmC* phase in the SmA* by applying an electric field (as well as by reducing the temperature, as above). When the field is applied parallel to the smectic layers, a polarization is induced in that direction. Since the polarization and the molecular tilt are coupled together, this causes the molecules to tilt with respect to the layer normal, i.e. the SmC* phase is induced.

This electroclinic effect is a promising one for display applications, as the electro-optic response is both analogue and fast. However, the effect is only really significant close to the SmA*–SmC* phase transition temperature, and therefore has a temperature dependence unsuitable for most applications.

1.5.2 SmC* field response

1.5.2.1 Simple bookshelf surface-stabilized ferroelectric liquid crystal device

Deep in the SmC* phase, where the soft mode response to an applied field is negligible (i.e. the tilt angle θ is essentially fixed), the electric field couples instead to the net spontaneous polarization that is present in the material. The director rotates around the smectic cone (changing the azimuthal angle, ϕ, via the 'Goldstone' mode) so that the polarization aligns with the applied field. If we assume that the device is sufficiently thin that the helical structure is suppressed, then there are two possible stable, uniform ground states, as shown in figure 1.9. In one state (say $\phi = 0$), the directors are on one side of the cone, and the polarization is pointing upwards, but in the other state the directors are on the other side of the cone ($\phi = \pi$) and the polarization is directed the opposite way. It is clear that the system is bistable, and that electric fields of opposite sign can be used to switch between the two

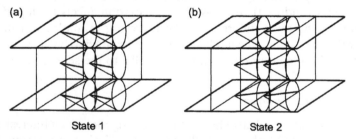

Figure 1.9. The surface-stabilized ferroelectric liquid crystal device (SSFLC) has two bistable ground states (a) and (b), with the director on each side of the smectic cone, parallel to the glass plates and uniform throughout the liquid crystal layer. Switching between the two states is achieved via an electric field applied perpendicular to the glass plates, which couples to the spontaneous polarization of the material.

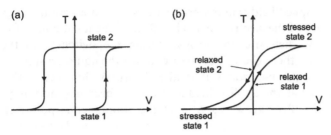

Figure 1.10. Electro-optic characteristics of typical SSFLC devices, when placed between crossed polarizers: in both bases, the bistability of the system causes thresholded behaviour and hysteresis. (a) An ideal bookshelf geometry, where the layers are formed perpendicular to the glass plates; one polarizer is aligned with the director of one of the stable ground states. (b) A more realistic chevron structure; one polarizer is aligned with the effective optic axis in one of the stressed states. The existence of a chevron structure reduces the optical contrast of the two relaxed states.

states. If crossed polarizers surround the device with one of them parallel to the optic axis of one of the stable states, then one of the two states will be black. The other state has its optic axis at an angle of 2θ to the polarizers, and therefore will transmit light according to a special case of equation (1.1):

$$T = \sin^2(4\theta) \sin^2\left(\frac{\pi \Delta n d}{\lambda}\right),$$

where $\chi = 2\theta$. A typical hysteresis loop for such a device is as shown in figure 1.10(a). For optimum contrast, θ should be 22.5°, a value typical of many SmC* materials. The thickness of the device should also be such that a half wave plate is achieved, that is:

$$d = \frac{\lambda}{2\Delta n}.$$

Because of the high birefringence of smectic liquid crystals, this often results in cell thickness of 2 μm or less.

In the operation of this bistable surface-stabilized ferroelectric liquid crystal (SSFLC) device [16], the switching between the two stable states is driven by the interaction between the applied electric field and the net spontaneous polarization. However, there is also an interaction between the applied field and an *induced* polarization. Unlike the electroclinic effect described above, this induced polarization is not due to the change in the smectic cone angle, but due to the polarization of the molecules themselves, i.e. it is a dielectric effect.

In a SmC* material, the axis of largest dielectric permittivity tends to be along the C_2 symmetry axis, i.e. along the direction of the spontaneous polarization. Therefore, in terms of the interaction with the dielectric anisotropy only, an applied field will favour both of the bistable states equally. The sign of the applied field is immaterial because the interaction energy depends on the square of the electric field. The effect of the interaction of the electric field with the dielectric anisotropy, during the switching of the bistable SSFLC device, therefore will be to discourage the liquid crystal from leaving whichever of the two states it is already in. At low fields, this effect is negligible, but as the applied field increases, it becomes more important (as the dielectric effect grows as E^2 and the spontaneous polarization effect grows like E), and eventually will begin to slow down the switching process. Therefore, as the applied voltage across the SSFLC increases, at first there is an increase observed in the speed of the device, but eventually the speed reaches a maximum value and the response of the device starts to slow down, as it is inhibited by the dielectric effect. The electric field at which the speed of the device is at its fastest is given by [17]

$$E_{min} = \frac{P_s}{\sqrt{3}\varepsilon_0 \Delta\varepsilon \sin^2 \theta}.$$

1.5.2.2   Effect of the chevron structure in SSFLCs

As described in section 1.3.3, the layer structure formed in SmC* devices where there is also a SmA* phase is not that of the simple bookshelf geometry assumed above, but instead forms a chevron structure. This has a significant impact on the optical properties of the SSFLC device. Assuming that the C2 structure is formed uniformly throughout the device, as it is lower in energy than the C1 structure (as described above in section 1.3.5), then the director structure in the two stable ground states is often assumed to be a uniformly twisting structure between the surface and chevron boundary conditions, as shown in figures 1.11(a) and (b). The effective optical tilt angle χ is determined by the angle of the cone intersection at the chevron interface, which in turn is determined by the layer tilt angle δ and the cone angle θ by [18]

$$\cos\chi = \frac{\cos\theta}{\cos\delta}.$$

In general, the effective optical tilt angle χ in the relaxed states tends to be about $\theta/3$ [19]. When an electric field is applied across the device, the motion of the directors around the smectic cone in order to maximize the alignment of the spontaneous polarizations with the electric field causes the effective optical tilt angle to increase. The condition of director continuity at the chevron interface means that the director in the centre of the cell is effectively pinned. The rest of the cell is free to distort in order to minimize the free energy, the result being a 'stressed state' as illustrated in figures 1.11(c) and (d). As the applied field increases, the director structure and hence the effective tilt angle saturate at a particular value. The actual value

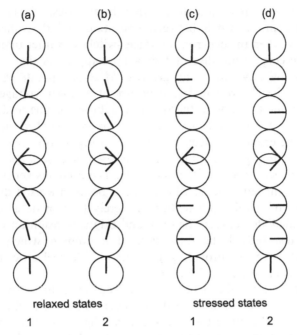

Figure 1.11. Illustration of the chevron states that exist at various points in the hysteresis curve of an SSFLC device: (a) and (b) show the two relaxed states that exist when there is no applied electric field; (c) and (d) illustrate the 'stressed states' that result from the application of a saturation electric field.

is determined again by the layer tilt angle, but is generally thought to be around the full cone angle θ. In any case, the exact value of the effective optical tilt angle is unimportant because the transmittance of the device is determined not only by the optical tilt angle but also by the effective birefringence, which is also dependent on the layer tilt angle.

A typical hysteresis curve that might result from a structure with a C2 chevron structure is shown in figure 1.10(b). The polarizers are generally oriented so that one of the stressed states is black. Therefore both the relaxed states are partially transmitting. Suppose that we start in relaxed state number 1, as shown in figure 1.11(a). As the voltage is increased, the directors away from the chevron interface reorient in order to align the spontaneous polarization with the applied field. The position of the directors at the chevron interface is unfavoured, and eventually the chevron interface will switch to the opposite state, so that stressed state number 2 is formed. As the applied field is reduced to zero, relaxation to the relaxed state number 2 occurs. When the electric field is reduced below zero, switching of the chevron interface across to the original position 1 occurs, and stressed state 1 occurs at high negative voltages. Then when the field is increased back towards zero, relaxed state 1 forms. In all cases, whether the applied field is sufficient to switch between one state and the other is entirely controlled by when the chevron interface switches. Rather than all parts of the cell behaving identically, in fact what happens is that the switching occurs by domain evolution and growth. At certain points within the cell, 'latching' of the chevron interface occurs at a lower voltage than other parts (either due to random, thermal fluctuations, or due to defects within the cell). These regions constitute domains of switched liquid crystal, which seed the chevron switching in neighbouring parts of the liquid crystal device. The regions grow until they have entirely coalesced and the whole of the device has switched into the new state. The domains generally take the form of boat-shaped regions, which always point in a fixed direction. The optical effect of the domain texture is merely to provide a transmittance that is intermediate between the dark and bright transmittances of the two bistable states. Indeed the formation of a domain structure has been suggested as a way of providing grey scale in ferroelectric liquid crystal (FLC) displays (see section 1.6.2.1).

It must be noted that when in a stressed state, the net polarization in each arm of the chevron in an SSFLC device is not parallel to the applied field because of the tilt of the layers: it is as aligned as it can be within the constraints that the layer structure provides. However, if a sufficiently large electric field is continually applied to the device, the net torque can be sufficient to reorient the layer structure into a bookshelf configuration. This effect has been studied by x-ray diffraction by Srajer *et al* [20]. However, in most cases where moderate electric fields are applied to an SSFLC device, the chevron structure can be assumed to remain intact.

Dielectric effects are also important in determining the switching properties of FLCs in the chevron structure. The dielectric anisotropy effects discussed in section 1.5.2.1 are now less important. However, the biaxiality introduced in section 1.2.2 is very important. Generally $\delta\varepsilon$ is positive and ε_{yy} is the largest dielectric permittivity. Therefore, application of a large enough field tends to stabilize the switched states through dielectric interactions. This effect is very important in optimizing addressing schemes for real FLC devices [21] (see section 1.6.3).

1.5.3 Distorted helix effect

When the pitch of the SmC* helix is less than the thickness of the device used, it is possible for the helical structure to exist in the ground state, in contrast to the surface-stabilized devices considered so far. The spontaneous polarization spirals through a total azimuthal angle of 360° in one pitch of the material, and thus averages to zero on a macroscopic scale. When a field is applied parallel to the smectic layers (perpendicular to the axis of the helix), it couples to the individual layer polarizations, causing a distortion of the helical structure. Eventually, at a critical field [22],

$$E_c = \frac{\pi^4 K}{4pP_s},$$

where K is the elastic constant of the material, p is the pitch of the helix and P_s is the spontaneous polarization, the helical structure is completely unwound, and all of the polarizations are aligned with the applied field. In so doing, the optic axis of the material has changed from pointing along the smectic layer normal to being on the side of the smectic cone, i.e. a tilt equal to the cone angle has been achieved. Exactly the same thing happens with the opposite sign of applied field, except that the optic axis tilts in the opposite direction. In this system, unlike the surface-stabilized system, there is no memory: as soon as the field is turned off, the helix rewinds and the optic axis returns to its original direction, along the smectic layer normal. The electro-optic effect is therefore an analogue, rather than a hysteretic one, rather like that of the electroclinic effect. Although the helix unwinding effect (or distorted helix effect, as it is more commonly known) is rather slower than the electroclinic effect, it is less susceptible to changes in temperature and shows a larger change in tilt angle (between $\pm\theta$), compared with only a few degrees for the electroclinic effect.

1.5.4   Twisted SmC* effect

Another analogue mode that can be achieved with ferroelectric liquid crystals occurs when the director structure formed in the ground state is not uniform in azimuthal angle throughout the cell gap (as in the bistable device described above), but twists continuously from $\phi = 0$ at one surface to $\phi = \pi$ at the other. This can occur when the interaction between the liquid crystal and the alignment layers is of a polar nature. This encourages the spontaneous polarization to point either into or out of both surfaces, and hence for the directors to occupy different positions on the smectic cone on the two surfaces. The director twists uniformly about the smectic cone between the two surfaces, causing an optical effect like that of a twisted nematic. When an electric field is applied across the cell, the director structure distorts, aligning the polarizations with the applied field. In the field-on states, the director is aligned with either of the polarizers, so that the device appears dark. The change in transmittance of the device with voltage occurs continuously and thresholdlessly. This analogue mode of FLCs was suggested and demonstrated by Patel [23]. In order to ensure the maximum useful swing in optic axis, a SmC* material with a 45° cone angle (which is quite common in materials that have a SmC* but no SmA* phase [24]) was used, and the substrates were rubbed in perpendicular directions, in order to ensure that the twisted state was achieved.

1.5.5 SmC$_A^*$ field response: tristable switching

In an AFLC device, there will initially exist a chevron structure, just as in an FLC device. However, the higher spontaneous polarizations that exist in AFLC materials ensure that applied fields will generally destroy the chevron structure in favour of a bookshelf structure [6]. In what follows, therefore, we can assume that the layers are formed perpendicular to the glass plates. Also like an FLC device, an AFLC device will retain a helical superstructure if the device is sufficiently thick, but becomes surface-stabilized if the thickness is less than roughly the pitch of the material.

In either case, if a sufficiently large electric field is applied parallel to the smectic layers, a field-induced transition into the ferroelectric state occurs, in order to maximize the alignment of the individual layer polarizations with the electric field. Likewise a field of the opposite sign will induce a transition to the opposite ferroelectric state. In both cases, when the field is removed, the liquid crystal returns to the original AFLC state. Thus an AFLC device can reach the two stable states of the ideal bookshelf SSFLC device under applied field. The switching between the antiferroelectric (AF) and ferroelectric (F) states is hysteretic (see later) and hence the electro-optic characteristics of the device involve a double hysteresis curve, as illustrated in figure 1.12 for the usual case where the polarizers are oriented parallel and perpendicular to the smectic layers, so that the AF state is black and the F states are equally bright. This arrangement has the advantage that the electro-optic characteristics are symmetric, which aids the design of DC balanced addressing schemes (see section 1.6.1).

1.5.5.1 *Pretransitional effect*

The use of AFLCs in displays has been hampered by the so-called pretransitional effect, which is the small increase in transmission of the AF state as the

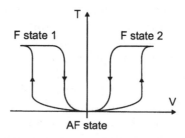

Figure 1.12. Illustration of the double hysteresis loop typical of an AFLC device, whether helical or surface stabilized. Like the FLC, the system is bistable, the cause of which is thought to be a significant quadrupolar component in the interlayer interaction. The pretransitional behaviour is quite different in the two cases of a helical and a surface-stabilized device.

applied voltage is increased, before the threshold for switching to the ferro-electric state is reached. This effect is believed to be caused by a small change in the antiferroelectric ordering in response to the applied field. In the ground state the polarizations in adjacent layers are antiparallel (or nearly anti-parallel in the case of a helical device) and hence cancel out. When a field is applied, the directors rotate around the cone in opposite directions in adjacent layers in order to cause a net polarization along the field. This has the effect of causing a small change in the net optical tilt angle away from the polarizers, and hence an amount of light leakage appears: the pretransitional effect [25].

1.5.5.2   Helical devices

In helical devices, the anti-phase rotation of the directors in adjacent layers described above (causing a net local polarization to exist) is accompanied by an in-phase rotation of the directors (that causes that polarization to rotate to align with the applied field). This in-phase rotation has the net effect of distorting the helical structure of the device. At a critical field, the helical structure unwinds completely to form a state in which the plane of the directors is approximately parallel to the applied field, because it is in this state that the induced polarization is along the direction of the applied field. The unwinding of the antiferroelectric helix is analogous to the unwinding of a cholesteric liquid crystal helix, as described by de Gennes [26].

Within the pretransitional regime of a helical AFLC device, therefore, there is a small change in the optical tilt angle (typically a few degrees) caused by an anti-phase director motion in adjacent layers, and a large change in the birefringence (typically 10%) caused by the in-phase motion or helix unwinding.

1.5.5.3   Surface stabilized devices

In surface-stabilized AFLC devices, the pretransitional behaviour is some-what different. If the pretransitional behaviour of the helical AFLC device is analogous to the field response of a cholesteric liquid crystal device, then that of a surface-stabilized AFLC is analogous to that of a planar-aligned nematic liquid crystal cell. In the ground state, the directors are in the plane of the glass surfaces. As the electric field is increased, first of all nothing happens to the director structure, and then at a critical field (the Freedericksz voltage) the structure begins to distort. The mechanism for the effect is exactly as for a helical device: an anti-phase motion of the directors brings about a net polarization in the plane of the directors; this polarization is then rotated towards the applied field by an in-phase motion of the directors in adjacent layers. With the helical device, this effect occurs to a greater or lesser extent depending on the exact position along the helical structure. In the surface-stabilized device the effect is dependent on the position across the gap between the glass plates. Just above the Freedericksz transition, the distortion is primarily in the centre of the device, and reduces to zero at the surfaces. As the voltage increases, the distortion grows outwards towards the glass plates until the unswitched region of the liquid crystal occupies only a thin region near the surfaces.

The electro-optic characteristics of a surface-stabilized AFLC device are therefore quite similar to those of a helical AFLC device, except that the pretransitional effect only occurs above a threshold voltage in the former device. Since the Freederickz effect gives rise to a threshold voltage and the AF to F transition depends on the applied field, it is possible to eliminate the pretransitional effect entirely by making a sufficiently thin device.

1.5.5.4 Hysteretic and thresholdless switching

The devices considered so far have exhibited thresholded switching, that is, there is hysteresis between the AF and both F states. The hysteresis has been hypothesized to originate both from the influence of the surfaces on the liquid crystal [27], and later due to the bulk interlayer interaction [28]. It is generally observed that the sharpness of the hysteresis curve increases with the thickness of the device used, indicating that the bistability is actually a bulk, not a surface phenomenon. This leads to the hypothesis that the interlayer interaction has a sufficiently high quadrupolar component that both the F and AF states are stable for a range of voltages, leading to bistability and hysteresis. In fact, in one material, the quadrupolar component of the interlayer interaction has been measured to be so high that the F state is stable for all electric fields, and it is only defect-seeded domain switching that causes the AF state to reappear once a device has been driven into the F state [29].

In some AFLC devices, however, a thresholded hysteresis loop is not observed. Instead, a thresholdless, analogue response, such as that shown in figure 1.13 occurs [30]. In these devices, it is believed [31] that the interaction of the bulk AFLC material with the polar surfaces of the alignment layer causes a twisted ferroelectric structure to occur, very similar to the

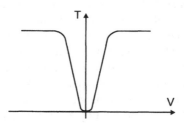

Figure 1.13. In some AFLC materials, the switching characteristics are not hysteretic, but instead are thresholdless or V-shaped in character, as illustrated here.

device suggested by Patel [23]. It is interesting to note that the formation of a twisted ferroelectric structure in preference to a twisted antiferroelectric structure may also be a consequence of a large quadrupolar component in the interlayer coupling.

1.6   Displays

One of the key areas of interest in the application of tilted smectic liquid crystals is in display technology. Demonstrator displays, of various types, using ferroelectric and antiferroelectric liquid crystal have been designed and built. Although some of these have been very impressive, commercialization has been somewhat limited.

There are two key features of tilted smectic liquid crystals which have made them attractive to display technologists attempting to build improved displays. One is the bistability in ferroelectric liquid crystals, which results from the internal smectic layer structure outlined above (section 1.5.2.1). In the case of antiferroelectric liquid crystals this is in fact a tristability, although only two of the states are normally arranged to be optically distinct in a display (section 1.5.5). This allows the materials to be potentially used in displays without the need for an active matrix circuit. The second key feature of ferroelectric and antiferroelectric liquid crystals is their relatively high switching speed, which allows them to be used in highly multiplexed displays. These features will be discussed further below.

1.6.1   Typical modes of operation

The most common mode of operation of a display using tilted smectic liquid crystals is to exploit the bistability/tristability of the switching processes in the materials. If a ferroelectric liquid crystal is used then a short electrical pulse can latch the material into one of the switched states. Using positive pulses to latch one way and negative pulses to latch the other way allows the latched state to be chosen. This can, in principle, be used to build a display with a relatively simple addressing scheme without the need for thin film transistors (TFTs) at each pixel.

It is reasonably easy to see how an individual pixel can be controlled by applying positive or negative pulses. However, the operation of a multiplexed display is somewhat more complicated. In this case electrodes can be placed on the substrates which are orthogonal to one another. The electrodes on one substrate then form rows, and the electrodes on the other substrate form columns (figure 1.14). Pixels are then formed at the intersections of the row and column electrodes. To control the switched state of the pixels on a given row, this row is 'selected' by applying an electrical signal to it, while all other row electrodes have no signal. Applying positive or negative

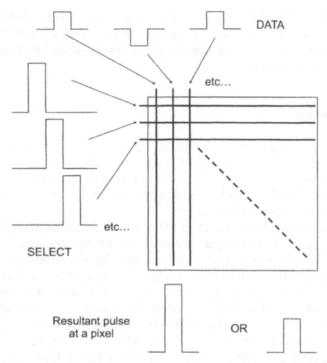

Figure 1.14. The principle of addressing in surface-stabilized ferroelectric liquid crystal display devices. The application of sequential select pulses to the rows, together with appropriate data pulses to the columns, results in switching pulses at the pixels. By controlling the amplitude of the data pulses the resulting pulse at a pixel may be either large enough to result in switching, or of insufficient size to cause switching.

'data' signals to the column electrodes will then either add to or subtract from the row signal for pixels in the chosen row. If the voltages are chosen appropriately then some will be sufficient to switch the chosen pixels. Once this has been done the 'select' signal can be applied to the next row, and the process repeated to switch chosen pixels on this row. Repeating the process for all rows allows an image to be written on the display device.

Because the process described above only controls the switching in one direction it is necessary to do one of two things in order to complete an image. Either a given row must be reset (normally into the dark state) before it is selected and data written to it, or two addressing frames must be used, one to switch chosen pixels towards the dark state and one to switch other chosen pixels towards the light state.

This then defines the basic operation of ferroelectric liquid crystal displays, although some displays use a more sophisticated addressing scheme, as further outlined below. The operation of antiferroelectric liquid crystal displays is very similar, but due to the tristability of the switching process an offset voltage is needed in order to access the hysteresis loop. This offset is normally reversed on alternate frames in order to ensure that d.c. balancing is achieved. That is, in order to ensure that there is no net d.c. voltage across a display device.

An alternative mode of operation for ferroelectric and antiferroelectric liquid crystal displays is to use them in an actively driven display system. Normally such systems are engineered as micro-displays on silicon backplanes and it is the high switching speed of the materials which is exploited. Because in this case the pixels are directly driven, it is possible to use either the bistability commonly present in ferroelectric liquid crystal systems or the deformed helix mode described in section 1.5.3. In this case it is the distortion in helix under applied field, and consequent changes in effective optic axis, which are exploited. Such micro-displays are of interest in projection display systems.

It is additionally possible to use tilted smectic materials as a direct replacement for twisted nematic materials in TFT-driven displays. For example, a thresholdless smectic structure can be used (see section 1.5.5.4) in place of a twisted nematic structure, with the possibility of fast switching and high quality viewing properties without the need for additional compensating films.

1.6.2   Grey scale in FLCs and AFLCs

Due to the bistable nature of the switching properties of ferroelectric liquid crystals the production of grey scale in displays is non-trivial. Two approaches have been taken to overcome this.

1.6.2.1   Spatial dithering

Provided the sections are small enough a display can be engineered with sub-pixels, which are not individually resolved by an observer. For example, in a display with pixels of the order of one hundred to a few hundred microns across these can be sub-divided into a number of smaller pixels. Switching of these sub-pixels can then lead to a grey scale effect. This can be achieved most usefully if the areas of the sub-pixels are arranged to form a binary series. Then, by switching different combinations of the sub-pixels a wide range of grey states can be formed.

An alternative way to achieve a similar effect is to exploit the switching process of the ferroelectric liquid crystal itself. Generally, the switching between states takes place through the formation and evolution of domains. Consequently if the switching signal is removed part way through the switching process a partially switched state exists. Because the individual domains are typically of the order of tens of microns across they are not resolved by a viewer, and a grey state is effectively created. However, accurate control of the partial switching is very difficult, and the method is not generally reliable.

1.6.2.2   Temporal dithering

An alternative way to achieve grey scale is to exploit the fast switching speed of the materials in order to switch any given pixel a number of times within a video frame. Then, for example, if a pixel is switched to the dark state for 50% of the time and the light state for 50% of the time a mid-grey will be observed. By controlling the portion of time for which a given pixel is switched light or dark a range of grey states can be achieved. By arranging the time slots during which a given row is multiply addressed to be in a binary progression it is again possible to have considerable control over the range of greys which can be produced.

1.6.3   Display addressing schemes

The simple display addressing scheme outlined above (section 1.6.1) is not necessarily the most useful in practice. Two schemes have been developed which have proved to be particularly interesting in practical displays.

The Joers–Alvey addressing scheme [32] exploits the combined influence of the spontaneous polarization in ferroelectric liquid crystals, and the dielectric properties of these materials. The switching is generally driven by the electric field interaction with the dipole, but resisted by the electric field interaction with the material's dielectric biaxiality. Therefore, because the interaction with the dipole is proportional to the field, but the interaction with the dielectric biaxiality is proportional to the field squared, a minimum switching time occurs for a particular field (section 1.5.2.1). If the field is further increased the dielectric term dominates and the switching becomes slower. It is therefore possible to design an addressing scheme where larger fields effectively prevent switching, and smaller fields cause it.

It turns out that such a scheme has certain advantages in practical display applications. It can be a very fast scheme, because it is operating at (or close to) the minimum switching time point for a given material. Additionally, the dielectric interaction can help to stabilize states in the device which show a large effective switching angle, giving a higher contrast display.

A second scheme which has shown considerable promise involves the use of pre-pulses to speed up the switching in a display. As outlined earlier the pulses used during switching must switch the display device either from the dark state into the light state, or from the light state into the dark state. However, by applying an appropriate pulse to any given line of a display at a time before it is selected for addressing, this situation can be improved. The so-called pre-pulse can be used to place the pixels on a given line into a state where they are effectively half way between the light and dark states. During the time of addressing it is then necessary to switch the pixels into either the light or dark state as required. This requires pulses of shorter duration and/or lower amplitude. The result is that the display can be operated faster and/or at lower drive voltages [33].

1.6.4 Typical FLC and AFLC displays

A number of flat panel demonstrator displays have been developed. However, that with the best visual appearance has been produced by Sharp, in collaboration with others [34]. This was a seventeen-inch diagonal full colour video rate high definition television resolution display. It was based on ferroelectric liquid crystals operating in the C2 alignment structure. It used an addressing scheme based on a modified form of the Joers–Alvey scheme. Grey scale was achieved through a combination of spatial dithering (by sub-pixels) and temporal dithering.

A number of liquid crystal on silicon-based micro-displays have been developed. The best known of these is that developed and produced by Displaytech. Their micro-display operates in a reflective mode and has fully integrated drive electronics. It is designed primarily for use in electronic viewfinders of digital camcorders.

A number of AFLC display prototypes have been developed [35]. For example, Denso have demonstrated a seventeen-inch full colour panel, which has wide viewing angles. Toshiba have used the materials in a thresholdless mode as a direct replacement for twisted nematic structures, and have demonstrated high quality displays using this technology.

1.7 Non-display applications

1.7.1 FLC spatial light modulators

The main building block of most non-display applications is the spatial light modulator (SLM). An SLM is an electro-optic device capable of modulating the intensity, phase or polarization of an optical wave front both in space as well as in time. Many SLMs are built around nematic liquid crystal technology; however, the fast switching of ferroelectric LC makes them the preferred material choice for high-speed SLMs. We will concentrate on this latter SLM class, which is based on the chiral smectic phase. A significant number of applications involving FLC/SLMs have successfully been demonstrated, mainly in the area of optical telecommunications and optical data processing. A representative selection of them will be outlined in the rest of the chapter.

1.7.1.1 Electrically addressed SLM

Electrically addressed SLMs [36, 37] come in both transmissive and reflective configurations. A typical transmissive geometry is shown in figure 1.15(a). A

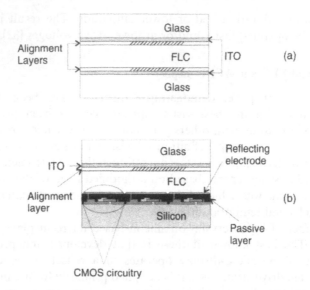

Figure 1.15. Electrically addressed spatial light modulator geometries: (a) transmissive with passive addressing; (b) reflective built on a silicon backplane with fast active addressing.

thin FLC layer of the order of 2 µm is sandwiched between alignment layers and is addressed by patterned ITO electrodes. The inherent FLC bistability allows for passive addressing. Reflective SLMs are built on silicon, as shown in figure 1.15(b), resulting in improved high-speed operation. A highly reflective electrode, usually made of Al, separates the FLC material from the VLSI silicon backplane, the latter accommodating the electronic circuitry. Dynamic or static random access memory addressing is used and a frame time set by the FLC switching speed in now permitted, leading to a display capability of thousands of frames per second. Typical pixel sizes are of the order of tens of microns and can go down to 7 µm with pixel counts varying from moderate 128×128 up to 1280×1024 or better.

The most obvious way of operating an SLM is as a binary intensity modulator. This is very similar to the optics of an SSFLC cell (see section 1.5.2.1) and they are simply summarized here once more. If the entrance and exit polarizers are placed parallel and aligned together with one of the stable FLC states then this can perform as the bright state. A perfectly dark state will be obtained if the switching angle is 45° (cone angle of 22.5°) together with a retardation $d\Delta n = \lambda/2$ (half wave plate).

Phase modulation is plausible as introduced by Broomfield [38]. The entrance polarizer is placed symmetrically with respect to the two stable FLC states, as shown in figure 1.16, with the exit polarizer crossed. Jones calculus [39] can verify that phase modulation of π radians takes place for the two FLC states, irrespectively of the FLC cone angle or the actual cell

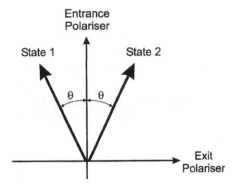

Figure 1.16. Arrangement of polarizer and analyser in an FLC/SLM intended for phase modulation.

retardation. However, there is a loss in light transmittance T as the latter is given by

$$T = \sin^2(2\theta)\sin^2\left(\frac{\pi d \Delta n}{\lambda}\right). \tag{1.2}$$

Optimized performance having maximum transmittance is achieved when the switching angle $2\theta = 90°$ (cone angle of $45°$) together with a retardation $d\Delta n = \lambda/2$.

1.7.1.2 Optically addressed SLM

In electrically addressed SLMs (EASLM) the displayed pattern is usually available in some digital format and it is uploaded by the interface electronics. Another class includes SLMs that can be addressed optically [36], allowing for the direct recording of an optically incident pattern. Optically addressed FLC/ SLM (OASLM) are made by combining a thin FLC layer with a photosensitive layer, usually of hydro-generated amorphous silicon (α-Si:H), as shown in figure 1.17.

A common voltage appears across the stack of the above layers via the ITO contacts. This voltage is divided between the FLC and α-Si:H layers and the actual voltage appearing across the FLC layer, which determines the FLC switching, is greatly dependent upon the incident light intensity of the write beam. In the absence of write light the α-Si:H exhibits very high resistivity limiting the voltage across the FLC layer and thus prohibiting switching. Sufficient write light lowers the α-Si:H resistivity, resulting in adequate voltage being dropped across the FLC. The local director orientation in the FLC film thus relates to the spatial intensity distribution of the write beam. This mode of operation is characterized as the photoconductor mode [40] and a simplified equivalent circuit is shown in figure 1.18(a). The

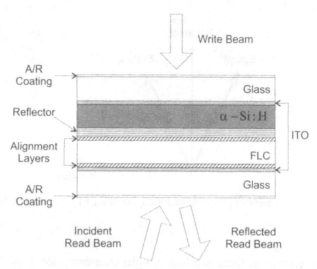

Figure 1.17. Optically addressed spatial light modulator combining a thin FLC layer and a photosensitive layer made of hydro-generated amorphous silicon (α-Si : H).

photosensitive layer is separated from the FLC layer by a dielectric mirror or a patterned metal mirror combined with a light-blocking layer to provide isolation between the write and read beams. ITO layers are not pixelated and the typical OASLM resolution is much higher than an EASLM.

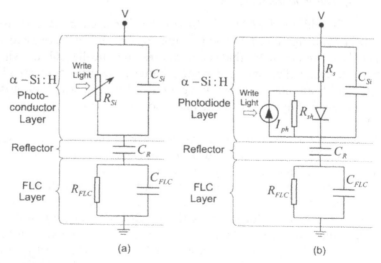

Figure 1.18. Equivalent electric circuits for OASLMs: (a) photoconductor mode; (b) photodiode mode. The FLC and α-Si : H layers as well as the reflector are modelled using lumped circuit elements.

OASLM operation can also be in photodiode mode [41]. In this mode the α-Si:H layer is fabricated as a p-i-n photodiode. Under forward biasing all the applied voltage drops across the FLC layer, switching it to, say, the OFF state. During this positive voltage period the FLC state is not affected by the occurrence of any write beam. Reversing the applied voltage polarity reverses the bias of the photodiode and now in the absence of any write light all voltage will drop across the α-Si:H layer. This will leave the FLC layer state unaffected. On the contrary, if write light illuminates the device during this negative voltage period the generated photocurrent will set the FLC to a negative voltage causing it to switch to the opposite (ON) state. Due to bistability (see section 1.3.2) this state will be retained until the next positive voltage period, which will effectively erase the recorded pattern returning the FLC layer back to the OFF state. A simplified equivalent circuit of the photodiode based OASLM is shown in figure 1.18(b).

Typical FLC-based OASLM characteristics include resolutions of the order of 20 to 100 line-pairs/mm, contrast ratios in excess of 200:1 together with sensitivities below 1 nW per pixel. Compared with other optically controlled switching devices, such as multiple quantum well (MQW) devices or nonlinear effect devices they demonstrate very high parallelism and sensitivity but also a much slower response time [42].

1.7.1.3 Basic SLM applications

Basic applications of EASLMs include fast shutters and polarization rotators. For instance, the simplest imaginable device can consist of a single electrically switched element and can be employed as a shutter/chopper with controllable duty-cycle [43]. It is very appealing for its size and lightweight, as well as having an absence of moving parts, as opposed to the classical motor-based chopper. A frequency range in the order of 10 kHz is common with a contrast ratio of around 1000:1. Single-element FLC devices can also provide fast polarization rotation/conversion in the visible and infrared.

OASLMs are well suited for various image-related operations [42]. Very basic ones include incoherent to coherent image conversion, where incoherent light is used for writing whereas reading and further processing is done with a coherent beam. Image amplification can also take place as a weak write beam and can be read with an intense coherent source. Wavelength conversion (i.e. different optical wavelengths are used for writing and reading) is also common: for instance, the write beam can be in the infrared with the read beam in the visible. OASLMs can also find applications in real-time holography: the interference of the object and reference beams is the write light on the photoconductor side and the hologram pattern is recorded by the OASLM. Hologram reconstruction follows by illuminating the pattern with a read-out beam.

1.7.2   Telecommunication applications

FLC technology has found many applications in the rapidly evolving area of optical telecommunication systems. One key element of these systems, which can be successfully implemented using FLC devices, is the all-optical free-space switch. A free-space switch interconnects a number of input and output ports in an optically transparent way and is capable of handling data irrespectively of their format or bit rate. The FLC switching speed is better suited to circuit-switched applications or network/traffic management operations rather than fast packet-switched applications. Other key telecom elements that can be implemented using FLC technology are wavelength filters for wavelength division multiplexing (WDM) and integrated wave-guide devices. We will examine some representative applications in the following sections.

1.7.2.1   2 × 2 fibre optic switch using FLC polarization switches

FLC 90° polarization rotators together with bulk optics components can be used to build 2 × 2 fibre optic switches [44]. The complete 2 × 2 switch will be polarization-insensitive and will comprise two separate polarization-dependent channels, for the *s* and *p* polarizations, respectively. For the sake of simplicity we will explain the concept of operation for the *s* channel by examining the polarization-dependent 2 × 2 switch of figure 1.19 and then we will generalize this idea to the polarization-insensitive switch structure.

The main optical components appearing in figure 1.19 are two FLC polarization rotators, PS1 and PS2, which in the ON state convert *s* to *p* light, a polarizing beam splitter (PBS) reflecting *s* light and passing through *p* light, a quarter waveplate (QWP), a half waveplate (HWP) and a total internal reflection (TIR) prism. The input state of polarization (ports 1 and 2) is *s* as set from the polarization maintaining fibres (PMF) and the *s* polarizer P1. When PS1 is OFF and PS2 is ON the switch is in the straight path mode (1 → 1′, 2 → 2′). The *s*-polarized beams of ports 1 and 2 are unaffected by PS1 and they are reflected by the PBS. They pass through the QWP, are reflected by the top mirror and then through the QWP for a second time. Their polarization is now changed to *p*, allowing them to go through the PBS. PS2 restores their polarization back to *s* and they finally reach their corresponding output ports (1′ and 2′). When PS1 is ON and PS2 is OFF the switch is in the exchange path mode (1 → 2′, 2 → 1′). The *s*-polarized beams of ports 1 and 2 change into *p* polarization after exiting PS1 and go through the PBS. In the TIR prism they are forced to exchange their positions and their polarization is further changed back to *s* as both of them had passed through the HWP. Their state of polarization is not affected by the PS2 and they finally reach the exit ports 2′ and 1′. The combination of PS2 and *s*-polarizer P2 helps maintain a low interchannel crosstalk. At the

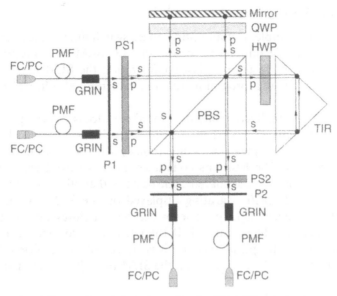

Figure 1.19. 2 × 2 fibre optic switch (*s* channel) using FLC 90° polarization rotators. PS1, PS2—FLC 90° polarization rotators; PBS—polarizing beam splitter; TIR—total internal reflection; QWP—quarter waveplate; HWP—half waveplate; GRIN—graded index lens; PMF—polarization maintaining fibre.

input and output ports graded index lenses (GRIN) are used to couple the light beams to and from the fibres.

A polarization-insensitive switch can be built by introducing a second *p*-polarization channel [44], similar to the one already described. Two more polarization beam splitters complete the switch. The first splits the arbitrary input polarization into the *s* and *p* components, which are directed to the *s* and *p* channels, respectively. The second PBS couples the outputs from the *s* and *p* channels to the switch output ports. This FLC-based fibre optic switch demonstrates an interchannel crosstalk of −34.1 dB, a switching speed of 35 μs and insertion losses around 6.9 dB, of which 4.4 dB are due to fibre coupling.

More complex switches can be implemented using a Banyan network with $\log_2(N)$ states where N is the number of inputs/outputs. Riza and Yuan [45] proposed a 4 × 4 switch based on a two-stage design with FLC polarization switches having four independent pixels. Also reconfigurable multi-wavelength add-drop filters can be designed by combining the basic FLC switch module with WDM demultiplexers to provide the wavelength separation for the input and add ports and multiplexers for the output and drop ports [45]. FLC polarization rotators offer negligible variation in their performance over the typical 40 nm WDM optical band centred at 1.55 μm.

1.7.2.2 *1 × N holographic optical switching*

One approach to implement an optical space switch interconnecting a single input to N possible outputs is to use an FLC/SLM as a programmable diffractive element. In the simplest configuration the SLM is patterned in stripes and it acts as a programmable one-dimensional diffraction grating [46].

A typical $4f$ architecture based on the above idea is shown in figure 1.20. A single mode input fibre is aligned along the switch optic axis whereas the output fibres make a linear (one-dimensional) array and are maintained in silicon V-grooves. The first lens collimates the wave front emerging from the input fibre and the collimated light beam is then diffracted at an angle dictated by the diffraction grating displayed on the SLM. A second lens focuses the diffracted light on one of the output fibres and as the fibre mode is matched light is coupled. If the nth fibre is located at a distance $s(n)$ from the switch optical axis and it is addressed by the first diffraction order of the grating G_n displayed on the SLM then the grating period $P(n)$ should be

$$P(n) = \frac{\lambda f}{s(n)} = v(n)\delta x. \tag{1.3}$$

In equation (1.3) $v(n)$ is the number of pixels per period of G_n, δx being the SLM resolution. A grating profile with maximized first-order diffraction efficiency is highly desirable: the binary nature of the FLC suggests a binary phase grating $(0, \pi)$ with 50% duty cycle. A maximum theoretical diffraction efficiency of $4/\pi^2$ (40.5%) is then possible. It is emphasized that light is symmetrically diffracted in both positive ($m = +1, +2, \ldots$) and negative ($m = -1, -2, \ldots$) diffracted modes making the 3 dB loss inevitable. This further means that in order to minimize the crosstalk due to symmetric orders, the output fibres must be placed to one side of the switch axis, say $s(n) > 0$.

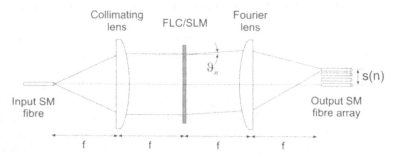

Figure 1.20. $1 \times N$ holographic optical switch based on an FLC/SLM one-dimensional diffraction grating in a transmissive $4f$ architecture.

Berthele *et al* [46] demonstrated the above concept for a 1 × 8 switch using a large tilt (≈ 45°) smectic C^* FLC material in a 256 × 1 SLM with 22 μm pixel size. The FLC/SLM provides truly polarization-insensitive operation (polarizers are not needed any more) as explained by Warr and Mears [47], with optimized performance when set to half waveplate at the wavelength of interest [38]. Polarization-insensitive operation is highly desirable as the state of polarization for light propagating down a fibre is unknown and can substantially vary over time. Typical switch insertion losses at 1.55 μm were found to be around 7.7 dB, of which 5.2 dB originate from the SLM losses (a diffraction efficiency of 27% instead of the theoretical limit of 40.5%) and 2.5 dB originate from the fibre-to-fibre losses. A switching speed of 400 μs was measured.

Crossland *et al* [48] reported a similar 1 × 8 polarization-insensitive switch in a 2*f* architecture using a reflective 540 × 1 LCOS FLC/SLM with a 20 μm pitch. A silica-on-quartz waveguide array was used for the input and output ports. Insertion losses were 16.9 dB at a wavelength of 1.55 μm together with a crosstalk of around 20 dB. The 3 dB optical bandwidth was found to be 60 nm.

It can be realized that the above one-dimensional topology does not scale well and it is estimated that it is hard to exceed a 1 × 16 switch design. Moving into a two-dimensional topology can substantially increase the switch capacity. In a two-dimensional topology the SLM displays a two-dimensional pattern and the output fibres are arranged into a two-dimensional array. Instead of a simple one-dimensional grating the SLM should now display a sophisticated computer-generated hologram (CGH), which provides steering of light to the selected output ports. Standard methods are available for calculating the CGH and they include the direct binary search (DBS) [49], genetic algorithms [50] and simulated annealing [51]. Further difficulties will relate to the accurate positioning of the output fibres in a two-dimensional array. For a transmissive SLM the 4*f* architecture is shown in figure 1.21(a) together with the more compact 2*f* version, figure 1.21(b), when a reflective SLM is used instead.

Warr and Mears [52] demonstrated a 4*f* system with 16 output fibres as shown in figure 1.21(a) at a wavelength of 780 nm. Using polarization-insensitive operation (without polarizers) there is always present some zero-order or undiffracted light. This light was sent to one of the output fibres, effectively making a 1 × 15 switch, and was used to facilitate system alignment and power monitoring. The output fibres were arranged in a 4 × 4 array with a 200 μm pitch on a laser-drilled kevlar plate. Irregularities in the array construction have been compensated by scanning a replay spot in the expected fibre positions and then selecting the CGH that maximized the light coupled out. A 320 × 320 transmissive FLC/SLM with 80 μm pitch and 5 μm intergap was used for displaying the binary phase CGH. Insertion losses are mainly attributed to the non-ideal phase modulation, which relates

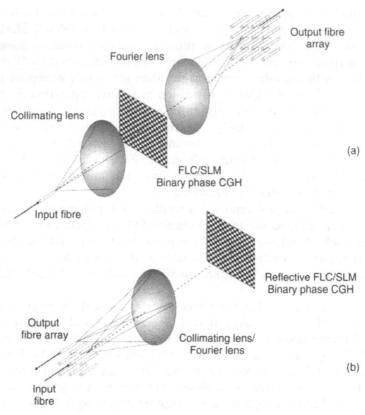

Figure 1.21. $1 \times N$ holographic optical switch based on a two-dimensional topology using a binary phase computer generated hologram (CGH): (a) $4f$ architecture employing a transmissive SLM; (b) $2f$ architecture employing a reflective SLM.

to the switching angle and the cell retardation as given in equation (1.2), the CGH diffraction efficiency, the various optical aberrations in the system and the fibre-to-fibre coupling losses. An average insertion loss of around 20 dB was measured for the various output fibres but this can be substantially reduced, as the phase efficiency of the FLC/SLM used was rather poor. Switching speed is limited by the FLC/SLM frame rate.

In a $2f$ system (figure 1.21(b)) the input fibre is aligned along the switch optical axis and it is combined with an isolator to prevent the zero-order diffracted light from interfering with the source. Output fibres are arranged in the same plane as the input fibre, resulting in a more compact and stable switch architecture, compared with the $4f$ case. A reflective FLC/SLM displays the CGH patterns, preferably using a fast actively addressed LCOS FLC/SLM device.

A holographic switch may suffer from high insertion losses but it also demonstrates some highly desirable characteristics: there are no moving parts, CGH are inherently redundant making the switch very robust to pixel failures, multiple output ports can be addressed by the appropriate CGH providing broadcast as well as routing operation and finally, crosstalk isolation can be over 30 dB. Also, as already mentioned, misalignments and fabrication errors can be corrected simply by changing the displayed CGH, possibly even during the switch lifetime. It is realistic to anticipate that insertion losses can be reduced to a level below 10 dB by optimizing the SLM and the system optical components.

A proposal for a large port count holographic switch came from Yamazaki and Fukushima [53] and was demonstrated in a 1×48 configuration. This holographic switch is optically controlled by an array of control-light sources. These coherent light sources are used to produce an interference pattern, which is recorded as a binary phase hologram on an FLC/SLM. The light to be switched is incident from the other side of the SLM; it is deflected in the direction determined by the holographic pattern and finally reaches the output port through a beam splitter.

An analysis of phase-only holograms with a number of phase levels $(2, 4, 8)$, SLM fill-factor issues, non-uniform illumination, an extensive study of crosstalk received by non-selected output fibres and fibre coupling issues can be found in recent publications by Tan *et al* [54, 55].

1.7.2.3 $N \times N$ holographic optical switching

The concept of $1 \times N$ holographic switching described so far can be extended to implement all optical $N \times N$ holographic switches with routing and/or multicast functionality. One possible switch architecture is shown in figure 1.22 [56]. Light emerging from each fibre belonging to the input array is collimated by an individual lenslet belonging to a lenslet array and hits a particular section of the FLC/SLM. This latter SLM section is only used by that single input fibre and displays the appropriate CGH deflecting the light accordingly. A single Fourier lens then focuses the deflected light from any input port into one or more fibres placed in the output array. The optical design of this architecture is demanding and high losses are expected, in particular for the outermost fibres, thus limiting the switch scalability and performance.

O'Brien *et al* [57] demonstrated this concept of $N \times N$ holographic switching for 16 input and 16 output ports at 850 nm, in a folded configuration using a 256×256 LCOS binary phase FLC/SLM. Input light came from a 4×4 array of vertical cavity surface emitting lasers (VCSEL) with an array of microlenses glued on top. The SLM was split into 16 sections and each one was used to address one of the inputs.

Other $N \times N$ holographic switch designs are based on two SLMs, each one still displaying an array of sub-holograms, as explained above. A 3×3 switch based on this two SLM approach was designed and constructed by

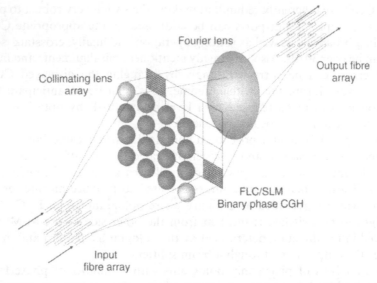

Collimating lens array

Fourier lens

Output fibre array

FLC/SLM Binary phase CGH

Input fibre array

Figure 1.22. $N \times N$ holographic optical switch based on a binary phase CGH. Every input fibre has its own lenslet and its light is deflected by a CGH displayed in a particular section of the SLM. A single Fourier lens refocuses the deflected beams on the output fibres.

Crossland *et al* [48]. High tilt FLC/SLMs without polarizers were used and a loss of 19.5 dB together with a crosstalk of −35.5 dB were measured at 1.55 μm.

1.7.2.4   *Fabry–Pérot type continuous tunable filters*

A Fabry–Pérot (FP) filter or resonator is an optical cavity made of a thin dielectric slab placed between two mirrors. The intensity transmittance has a wavelength dependence given by Born and Wolf [3]:

$$\frac{I^{(t)}}{I^{(i)}} = \frac{T^2}{(1 - R)^2 + 4R\sin^2(\delta/2)}. \tag{1.4}$$

In equation (1.4) R and T are the mirror reflectivity and transmissivity, respectively, and the phase difference δ is equal to

$$\delta = \frac{4\pi}{\lambda_0} nd\cos(\vartheta') + \varphi. \tag{1.5}$$

In equation (1.5) n is the slab refractive index, ϑ' is the angle of propagation with respect to the mirror normal inside the slab and φ is the additional phase shift introduced on reflection. Equations (1.4) and (1.5) suggest that resonance peaks will occur at wavelengths satisfying the condition

$\delta = 2m\pi$, m being an integer. Wavelength tuning can be obtained by varying the slab refractive index and this can be achieved by utilizing a suitable liquid crystal electro-optic effect. In this case the filter is active and the liquid crystal acts as the cavity active material.

Initial efforts at making liquid crystal based FP filters employed nematic technology [58]. Although nematic FP filters allow for broad tunability, their switching speed is rather poor ranging from tens to hundreds of milliseconds. For high-speed filter operation a ferroelectric LC material should be used instead.

A smectic C^* liquid crystal with a 45° cone angle (see section 1.5.4) was used in a two-wavelength filter (binary wavelength switch) [59]. In such material the slow and fast optical axes are interchanged by reversing the polarity of an externally applied field. For randomly-polarized light at two resonant wavelengths, denoted λ_1 and λ_2, the transmitted light will be polarized along one direction for λ_1 and at a perpendicular direction for λ_2. Reversing the electric field across the FLC filter interchanges the slow and fast axes and as a result the polarization directions for λ_1 and λ_2 are interchanged. The FLC FP device was constructed by placing dielectric mirrors on top of ITO-coated glass plates. An alignment layer was deposited on top of the mirrors and the two plates were rubbed in orthogonal directions in order to ensure symmetric switching between the two states. Experimental results for a 6.5 μm thick device filled with CS 2004 FLC material at an optical wavelength of 1.55 μm demonstrated an extinction ratio of 100:1 together with a switching time of around 2 ms [59].

Wavelength division multiplexing (WDM) optical communication systems are recognized as the key technology to explore the immense fibre bandwidth. One of the key components in implementing this technology is a tunable filter at the receiving end. An analogue smectic C^* FLC tunable filter in an FP cavity capable of providing fast continuous tuning would be highly desirable for WDM applications [60]. Parallel rubbed cells resulting in chevron structures (see section 1.3.3) or anti-parallel rubbed cells resulting in quasi-bookshelf structures, together with a sufficient device thickness can provide analogue refractive index modulation. Experimental studies were carried out at the 1.55 μm band for a 16 μm thick filter filled with CS 1014. Dielectric mirrors were used for the FP cavity and a finesse of 62 was measured for the filled device. The anti-parallel rubbed FLC FP filter demonstrated a continuous tuning range of 14 nm with a FWHM of 0.84 nm. The above tuning range is obtained by varying the external electric field in the range 0–4 V/μm. Insertion losses are very low at just under 0.7 dB combined with a switching speed of 400 μs.

A substantial improvement in switching speed can be gained by using chiral smectic A^* electroclinic liquid crystals as the active cavity material (see section 1.5.1). Electroclinic LC filters operating at 1.55 μm can potentially deliver a tuning range of around 30 nm (being compatible with the

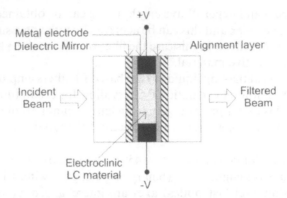

Figure 1.23. Fabry–Pérot electroclinic LC tunable filter. The external voltage, which tunes the filter, is applied in the lateral direction (i.e. perpendicular to light propagation).

bandwidth of erbium-doped fibre amplifiers) together with a switching speed of less than 10 µs [61]. The proposed electroclinic filter structure by Sneh *et al* [61] and Sneh and Johnson [62] has a noticeable difference compared with the filters already discussed: the external voltage is now applied in the lateral direction as shown in figure 1.23, as opposed to the typical case where the voltage is applied across the filter. It follows from equation (1.5) that the tuning range $\delta\lambda$ is related to the refractive index modulation through

$$\delta\lambda = \frac{\lambda}{n}\delta n. \qquad (1.6)$$

To obtain a useful tuning range, having in mind that practical electroclinic reorientation angles are in the order of 8–10°, one must introduce a significant pretilt for the LC molecules at the filter surfaces. Obliquely evaporated SiO_x can be used to produce a pretilt of around 30°, which in turn leads to sufficient index modulation. Experimental results were reported for a FP electroclinic LC filter for visible light, achieving a tuning range of 10.5 nm for an applied electric field in the range -8 V/µm to $+8$ V/µm. The switching speed was measured to be around 9 µs.

Sneh and Johnson [62] also demonstrated a compact version of the above device in a fibre configuration, the fibre FP electroclinic LC filter. Two single mode fibres at 1.55 µm were inserted into precision ferrules and the endfaces were polished and coated with broadband dielectric mirrors. In order to reduce large insertion losses due to fibre mode diffraction inside the cavity a waveguiding fibre piece was attached to one of the fibre endfaces. Oblique SiO_x evaporation was used for surface alignment at the fibre endfaces, and as previously a lateral electrode structure was introduced. A 17 µm cavity was formed (of which 10 µm is the waveguiding piece) with a finesse of 70 and a bandwidth of 0.68 nm. A tuning range of around 13 nm was achieved for electric fields up to 5 V/µm. An insertion loss of about

7 dB was measured. It is theoretically predicted that insertion losses can be drastically reduced for higher finesse mirrors. Switching speeds even down to 6 μs at 35 °C were reported.

The superior switching characteristics for the electroclinic FP filters come at a cost, and this is the high operating voltage. In many cases it is particularly attractive to use much lower voltages, for instance compatible with liquid crystal on silicon (LCOS) VLSI technology. Deformed helix ferroelectric (DHF, see section 1.5.3) liquid crystals can be used as the active cavity material in FP tunable filters operating at low voltage and high speed [63]. The helical structure is oriented perpendicular to the cell normal (direction of light propagation) and due to the short sub-wavelength pitch presents an effective medium to the incoming light. An applied voltage deforms the helical structure resulting in a modulation of the refractive indices. The typical driving electric field is around 1 V/μm and a tuning range of around 10 nm has been demonstrated for visible light. Switching speed is of the order of hundreds of microseconds, and although much slower compared to the electroclinic filters, it is still faster compared to the high tilt FLC devices.

1.7.2.5 Digitally tunable optical filters

The wavelength dispersion of CGH can be usefully exploited to make digitally tunable optical filters [64–66]. The key element will be an FLC/SLM acting as a programmable-phase-only grating by displaying patterns of different spatial period. A filter of this kind implemented as a 4*f* system is depicted in figure 1.24. Light from a single mode input fibre is collimated by the first lens and is then diffracted by the binary phase SLM displaying a one-dimensional grating, as previously explained in the holographic switch section. As the typical FLC pixel size is quite large with respect to the optical wavelength, it will significantly restrict the tuning resolution for practical

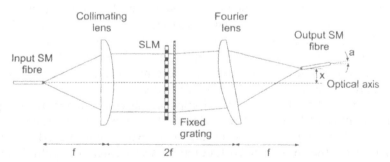

Figure 1.24. Digitally tunable optical filter using a binary phase FLC/SLM and a fixed diffraction grating. Based on the pattern displayed on the SLM single or multiple wavelength filtering is possible.

telecom applications. This can be resolved by placing adjacent to the SLM a fixed phase grating with a much higher spatial frequency (i.e. smaller grating period). In this arrangement the fixed grating further diffracts the light coming out of the SLM. A further lens converts the angular wavelength separation into positional separation. Depending on the grating displayed on the SLM, a particular wavelength will be coupled into a fixed output fibre placed at the second lens focal plane, the fibre being inclined a few degrees with respect to the system axis in order to optimize the coupling efficiency. Using a small angle approximation it can be shown that the resonant wavelength for first-order diffraction is given by

$$\lambda \cong \frac{x}{f\left(\dfrac{1}{d_{\text{SLM}}} + \dfrac{1}{d_{\text{fixed}}}\right)}. \tag{1.7}$$

In equation (1.7) x is the output fibre offset, d_{fixed} is the period of the fixed grating and d_{SLM} is the spatial period of the reconfigurable grating displayed on the SLM. The smallest value addressable is $d_{\text{SLM}} = 2D$, D being the SLM pixel size.

The above filter based on the two grating combination was proposed and characterized by Parker and Mears [65]. A low resolution 128×128 transmissive SLM with 165 μm pixel size was combined with a fixed 18 μm photoresist grating. A wide tuning range of 82 nm centred at 1.55 μm was possible, together with discrete tuning steps of around 1.3 nm and a FWHM of 2 nm. Wavelength isolation was 20 dB at 3 nm from the central passband wavelength but insertion losses were quite high (22.8 dB), mainly due to the small FLC switching angle. A distinctive feature of the digitally tunable filter, as opposed to the analogue counterpart of the previous section, is the ability to filter multiple wavelengths simultaneously by displaying the appropriate pattern. FLC bistability ensures that the filter is still functional even if electrical power fails; however, reconfiguration is no longer possible. More compact $2f$ filter designs were also proposed and assessed [66], which can be based on the combination of a transmissive FLC with an inclined fixed reflective grating or a transmissive fixed grating with a silicon backplane (LCOS) FLC. A filter architecture without a fixed grating is very unlikely, as it will require an unrealistic number of pixels together with an extremely small pixel size in order to meet WDM telecommunication systems specifications requiring a resolution of 0.8 nm or better.

A tunable fibre laser for WDM applications can be made by incorporating the filter of figure 1.24 in a closed fibre loop with an erbium-doped fibre amplifier (EDFA) to provide gain, as shown in figure 1.25. The pattern displayed on the SLM can be tailored to select single or multiple lasing wavelengths on demand. CW operation was demonstrated for the whole erbium window of 38 nm centred at 1.55 μm using tuning steps of 1.3 nm, as expected from the filter characteristics [66]. Lasing linewidth was

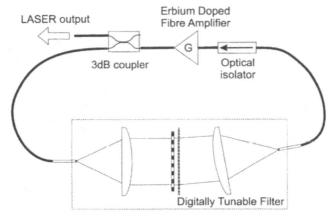

Figure 1.25. Tunable fibre laser based on the digitally tunable filter of figure 24. The filter is incorporated in a closed fibre loop with an erbium-doped fibre amplifier (EDFA) as the gain element.

measured to be in the order of 3 kHz. Other filter applications cover active channel management such as equalization or amplification in a WDM system [67].

1.7.2.6 FLC-based optical waveguides, switches and modulators

Up to this point we have discussed the implementation of optical switches using free-space optics and FLC/SLMs. FLC materials can also be used to make integrated optical waveguide devices that can function as optical switches or intensity modulators. FLCs have effective electro-optic co-efficients at least an order of magnitude greater than other materials used in integrated optics (such as $LiNbO_3$ or GaAs operating with the Pockels effect) and together with the available low-cost process technology and silicon backplane compatibility this makes them attractive for very compact low-cost integrated optical devices. However, FLCs have certain drawbacks, the most noticeable being the slow response, of the order of 20 μs as opposed to the Pockels or Stark effect response times of the order of 10 ns. Light scattering losses due to fluctuations in the molecular alignment is another limiting factor, but clearly FLCs provide substantial improvement over the nematics as their higher degree of ordering reduces losses to around 2 dB/cm, which does not exclude them even from being used as waveguide core materials. Low-cost FLC-based integrated optical devices with very short interaction lengths are feasible, provided their switching speed is considered acceptable.

FLC integrated waveguide devices typically use a surface-stabilized geometry (SSFLC) with the LC director lying on the plane of light

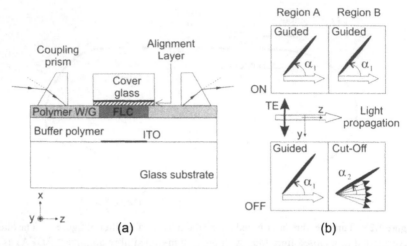

Figure 1.26. Waveguide electro-optic modulator using an FLC guiding layer: (a) device schematic; (b) principle of cut-off type of operation. The device can modulate only TE modes.

propagation, and the FLC functioning as a uniform switchable uniaxial layer. Two major classes of devices have been proposed and demonstrated: the first one uses the FLC as the guiding layer whereas the second uses the FLC as a cladding layer. Walker *et al* [68] theoretically analysed ferroelectric waveguide modulators of the deflection and cut-off type with the FLC being itself the guiding layer, as shown in figure 1.26(a). Figure 1.26(b) depicts the operation principle for the cut-off modulator. In the ON state the optic axis in regions A and B is oriented along the same direction selected to support at least one guided mode. In the OFF state the optic axis in region B is oriented so that all modes are now below cut-off. Light is no longer guided in region B and will be lost in radiation modes. Hermann *et al* [69] prepared and characterized an electro-optic modulator (± 30 V) for operation in the visible based on the concept of figure 1.26. Polymer waveguides made by photochemical cross-linking were used and the smectic layer normal was arranged at $\alpha = 44°$ with respect to the light propagation direction. In the ON state the director was aligned at $\alpha_1 = a + \theta = 67°$ ($\theta = 23°$ is the cone angle) whereas in the OFF state it was aligned at $\alpha_2 = a - \theta = 21°$. Therefore, in the ON state light experiences a higher effective index in the FLC layer and guidance is attained, with the OFF state providing a lower refractive index leading to cut-off. It should be clear that the above discussion explicitly refers to TE waveguide modes, as TM modes are insensitive to the FLC anisotropy changes.

The second class of devices employs the FLC material as a cladding layer and thus greatly reduces Rayleigh scattering losses. Ozaki *et al* [70]

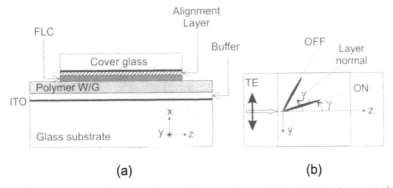

Figure 1.27. Waveguide electro-optic modulator using an FLC cladding layer: (a) device cross-section; (b) top-view of FLC orientation. Binary modulation is possible using the SmC* phase and analogue modulation is also possible using the SmA phase and the electroclinic effect.

demonstrated an electro-optic modulator based on a polymer waveguide with an FLC cladding layer. A cross-section is shown in figure 1.27(a) together with a top-view in figure 1.27(b). Light polarized parallel to the layers (TE) experiences an effective index

$$n_{\text{eff}} = \frac{n_{\parallel} n_{\perp}}{\sqrt{n_{\perp}^2 \sin^2(\gamma) + n_{\parallel}^2 \cos^2(\gamma)}}, \qquad (1.8)$$

with γ being the angle between the light propagation direction and the LC director. The smectic layer normal is arranged along the critical angle γ_c, which corresponds to an effective index n_{eff} equal to the polymer waveguide index n_p. In the OFF state the director is at an angle γ exceeding the critical angle ($\gamma > \gamma_c$) and the total internal reflection condition at the waveguide/ FLC interface is violated, resulting in loss of guidance. In the ON state $\gamma < \gamma_c$, restoring the conditions for total internal reflection. This device can operate in the S_c^* phase, providing binary modulation with a contrast ratio of around 40. More interestingly it can also operate in the S_A phase, using the very fast electroclinic effect and provide analogue electro-optic modulation, as shown by Ozaki et al [70].

An integrated optical waveguide switch can be implemented by embedding a thin layer of SSFLC between two planar waveguide as illustrated in figure 1.28. This is in essence a vertical directional coupler with one FLC orientation corresponding to the switch bar state and the other orientation corresponding to the switch cross state. An initial proposal of this device came from Clark and Handschy [71] using ion-diffusion multimode waveguides and a shear-aligned FLC layer. The smectic layer normal makes an angle θ equal to the cone angle with the light propagation

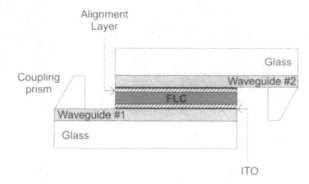

Figure 1.28. Integrated optical waveguide switch based on a SSFLC layer. Based on the FLC orientation the two waveguides can be decoupled (bar state) or strongly coupled (cross state).

direction. This sets the light propagation direction along the FLC optic axis for one voltage polarity and thus a TE mode will sense the lowest (ordinary) refractive index. Under this condition the two waveguides are decoupled and the switch is in the bar state. The opposite voltage polarity will increase the effective refractive index presented by the FLC layer, as predicted by equation (1.8), and provide strong coupling of light into the FLC layer and into the other waveguide setting the switch in the cross state. In this early proposal switching voltages were excessive, being of the order of ± 1 kV. D'Alessandro *et al* [72] demonstrated a similar switch device using single mode waveguides made by ion exchange on BK7 glass and operating with 20 V pulses. Prototype characterization revealed a maximum extinction ratio of 15 dB combined with a switching speed of about 300 μs. Asquini and d'Alessandro [73] used the beam propagation method to optimize the parameters of this optical waveguide switch for operation at 1.55 μm. The influence of the FLC orientation angles and the refractive indices of the waveguides and buffer layers were theoretically studied. Simulation revealed the possibility of optimized devices with extinction ratios in excess of 50 dB, losses better than 1 dB and very short coupling lengths of 175 μm.

1.7.3 Optical data processing applications

A light beam carries information with very high throughput, offered by the inherent parallelism of light in free space. Processing optical signals can prove extremely effective for inherently two-dimensional data, for instance optical images, compared with the more conventional VLSI processing of electrical signals. Various image processing applications, including logic operations, moving object extraction and pattern recognition, have been successfully demonstrated using the FLC/SLM technology. We will briefly discuss some of them in the following sections.

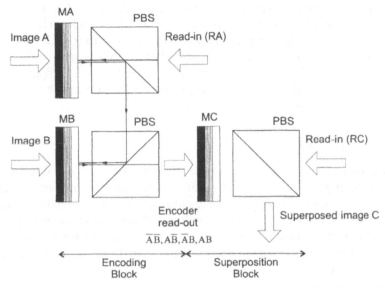

Figure 1.29. Optical processor for performing logical (Boolean) operations between binary images A and B. The encoder consists of OASLMs MA and MC and can output $\overline{A}\,\overline{B}$, $A\overline{B}$, $\overline{A}B$, AB, which are selectively latched by the superposition block consisting of OASLM MC.

1.7.3.1 Optical parallel processing of binary images

Optical parallel processing of binary images is possible with OASLMs as demonstrated by Fukushima and Kurokawa [74]. In particular, all Boolean operations between two binary images can be performed in real-time using three cascaded bipolar-operational OASLMs (abbreviated as B-OASLM), which are bistable and capable of reading-out the images in positive or negative mode, depending on the polarity of applied electric pulses [40]. A schematic of the optical processor is shown in figure 1.29. The system comprises two blocks, a time-domain encoder block using two B-OASLMs (MA and MB) and a superposition block involving the third B-OASLM (MC). The two input images (A and B) are incident on the photoconductor side of MA and MB. By changing the polarity of the erase and write pulses applied on MA and MB over the four possible combinations corresponding to positive and negative operation mode, the encoder output will consist of the logic operations $\overline{A}\,\overline{B}$, $A\overline{B}$, $\overline{A}B$, AB in sequence. The superposition block consisting of MC will latch some of the above four encoded images, depending on the electric pulses applied to MC. In general, if T_i, $i = 0, 1, 2, 3$, are regarded as the logic values of the electric pulses to MC then the output image will be equal to

$$C = \overline{A}\,\overline{B}T_0 + A\overline{B}T_1 + \overline{A}BT_2 + ABT_3.$$

For instance, the exclusive OR (XOR) operation requires only the latching of $A\overline{B}$ and $\overline{A}B$ by MC and can be executed in two clock-cycles. However, the complete set of Boolean operations will require up to four cycles. After the completion of these four cycles a read-in beam (RC) is sent to the super-position block and the superposed image C is read-out. A compact version of this system was demonstrated using camera lenses for image input and a laser diode as a read-out source.

1.7.3.2   Optical correlation

Correlation provides a measure of the similarity between two functions (or images). Being able to perform correlation using light beams can potentially lead to very high speedups. Applications of optical correlation include, among others, optical character recognition, object identification, fingerprint recognition and optical inspection. If $s(x,y)$ is an unknown input image and $r(x,y)$ is a reference image then correlation is mathematically expressed by the operation

$$s(x,y)^*r(x,y) = FT\{S(u,v)R^*(u,v)\}, \qquad (1.9)$$

where $S(u,v)$ and $R(u,v)$ are the Fourier transforms of $s(x,y)$ and $r(x,y)$, respectively. There are two common optical correlator architectures: the classical VanderLugt type correlator and the joint transform correlator (JTC). Both architectures have been successfully implemented using FLC technology and they will be briefly discussed.

In a VanderLugt correlator [75–77], the Fourier transform of the input image $s(x,y)$ is performed optically using a lens while the Fourier transform of the reference image (referred to as the filter) is typically calculated by a computer. A generic schematic of a $4f$ correlator is shown in figure 1.30. SLM1 displays the input image in binary-amplitude mode and it is Fourier transformed by lens L1. Reference image is Fourier transformed off-line and $R^*(u,v)$, after discarding the amplitude information, is binarized in order to produce a binary phase-only filter (BPOF). The BPOF is displayed on SLM2, which is used in binary-phase mode. The BPOF is simply obtained by thresholding the phase according to the following (although not unique) rule:

$$\text{BPOF} = \begin{cases} 0 & \text{Re}[R(u,v)] > 0 \\ \pi & \text{elsewhere.} \end{cases}$$

Light exiting SLM2 is the product of the input image Fourier transform and the filter, and is further Fourier transformed by lens L2 completing the correlation operation given by equation (1.9). Intensity peaks appearing in the exit plane (correlation plane) reveal the position of the reference image found in the input image. Using a BPOF introduces some degradation in the correlator performance due to the forced binary phase; however, these

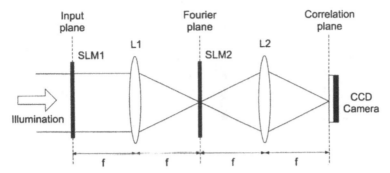

Input plane   Fourier plane   Correlation plane

L1   SLM1   SLM2   L2

Illumination   CCD Camera

f   f   f   f

Figure 1.30. Generic VanderLugt optical correlator with a binary phase only filter (BPOF) placed in the Fourier plane. Correlation peaks are recorded on a CCD camera.

penalties are far outbalanced by the convenience offered employing SLM technology. The generic correlator of figure 1.30 is usually quite impractical in terms of physical dimensions, and practical system implementations usually involve a magnification stage and/or silicon backplane SLMs.

Correlation is inherently shift-invariant but this is not true for other highly desirable forms of invariance for practical applications, such as scale or rotation invariance. Recently various techniques around simulated annealing were introduced to add invariance. This is done by combining a set of references images, for instance of varying size, in a filter containing features for all the original references. Wilkinson *et al* [75] demonstrated scale-invariance optical correlation using a BPOF in real-time road-sign recognition. Typical difference in correlation peaks was around 6–7 dB with a small variation in correlation-peak height of 15% for an area scaling between 1.0 and 2.4: Keryer *et al* [77] generalized the VanderLugt BPOF correlator in order to allow for multi-channel operation and demonstrated four-channel spatial multiplexing.

In the joint transform correlator (JTC) [76–80], the input image and the reference are displayed side-by-side and separated by a distance of $2a$ on an input-plane electrically addressed FLC/SLM. The input is mathematically written as $r(x - a, y) + s(x + a, y)$ and the intensity pattern $I(u, v)$ of the joint Fourier transform is given by

$$I(u, v) = |R(u, v)|^2 + |S(u, v)|^2 + R(u, v)S^*(u, v)\exp(-2\pi j2au)$$
$$+ R^*(u, v)S(u, v)\exp(+2\pi j2au).$$

Clearly, if a second Fourier transform is performed on the intensity pattern $I(u, v)$ it will contain on-axis the autocorrelation of the input and reference images (terms $|S|^2$, $|R|^2$) as well as cross-correlation terms placed off-axis at coordinates $x = 2a$ and $x = -2a$. A generic schematic of a JTC is shown in figure 1.31. Lens L1 performs the joint Fourier transform of the input SLM and the intensity pattern $I(u, v)$ (or joint power spectrum) is

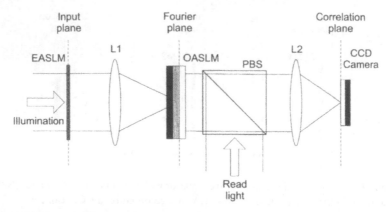

Figure 1.31. Generic joint transform correlator. Input and reference are displayed side-by-side on the EASLM. The OASLM implements a controlled nonlinearity in the Fourier plane.

written on the FLC/OASLM placed at the lens back focal plane. A second light beam reads the OASLM and it is Fourier transformed by lens L2 and imaged on the correlation plane, which is a CCD camera. In the correlation plane one observes the d.c. autocorrelation term together with two symmetric peaks, which are the cross-correlations between input and reference, as explained above. The OASLM is very well suited for implementing controlled nonlinear processing in the Fourier plane that can lead to sharp correlation peaks and improved SNR. OASLM nonlinearity follows a kth law and by changing the voltage-driving parameters for the OASLM this nonlinear response can be modified. Wilkinson *et al* [76] and Guibert *et al* [79] developed a JTC for real-time road-sign recognition. Scale invariance was successfully addressed by the linear combination of several reference views with appropriate weight factors obtained by a simulated annealing algorithm. Petillot *et al* [80] developed a JTC prototype for fingerprint recognition with built-in rotation invariance based on a similar linear combination/simulated annealing technique: the inherently poor rotation invariance of the JTC was extended from $\pm 2°$ to nearly $\pm 10°$. Multi-channel spatial multiplexing in JTC was also examined [77], and a similar increase in capacity was obtained as in the case of VanderLugt BPOF correlators.

Comparing the two correlator-architectures, the VanderLugt BPOF correlator has the filter located at the Fourier plane with the filter accommodating several different features to allow for invariance. Binary phase SLMs lead to sharp peaks with good signal-to-noise ratio and finally, the separation of the input and filter planes results in a single correlation making detection easier. On the other hand a JTC is a more compact and robust architecture with less optical alignment problems. However, displaying input image and reference side-by-side reduces the spatial bandwidth product (SBWP) and also decision making in the correlation plane is now more difficult due to the occurrence of the symmetric correlation peaks.

1.7.3.3   Optical neural networks, ferroelectric liquid crystals and smart pixels

The inherent parallelism offered by optical systems make them a very strong candidate for overcoming the interconnection bottlenecks found in high-complexity neural networks. Ferroelectric LC also prove successful in optical neural networks, as demonstrated by various research (see for instance [81–85]). A key element allowing for the implementation of these optical processing systems is the 'smart pixel', which is an element capable of processing information carried by light and subsequently modulating light to be further processed [83, 85–88]. In essence smart pixels are a generalization of the SLM and each of them may combine memory, sensing elements (one or more photodetectors), intra-pixel processing capabilities (such as amplification, thresholding, summation, logical operations, etc.), inter-pixel communications (usually with neighbouring smart pixels) and optical output provided typically by an FLC SLM for further connectivity. Smart pixels are arranged in two-dimensional arrays and are built on a VLSI silicon backplane.

1.7.4   Other applications

1.7.4.1   Photonic delay lines for phased-array antenna systems

The implementation of variable photonic delay lines (PDL) using FLC devices has recently been proposed and demonstrated [81–90]. PDL are attracting increasing interest in microwave engineering and in particular in phased-array antenna applications, as they offer wide-band operation together with high immunity to electromagnetic interference and electromagnetic pulses. In a PDL the delay is introduced by forcing the optical signal (which is obtained by modulating a light source with the RF signal) to follow different path lengths bypassing the straight path. Cascading delay blocks, which can be separately addressed (thus turning their individual delay on/off), can insert a variable delay. A possible delay block (termed as bit) using feed forward is shown in figure 1.32. Operation is based on polarization switching implemented by FLC polarization rotators. As already explained in a previous section, FLC polarization rotators change the incoming polarization (say form s to p) when set to ON and leave it unaffected when set to OFF. If FLC1 is ON and FLC2 is OFF incoming s light changes into p, passes through the PBS1 and PBS2, remains unaffected by FLC2 and it is transmitted through the final fixed polarizer P2. This is referred to as the straight path. Reversing the setting of FLC1 (OFF) and FLC2 (ON) introduces the bypass and as a consequence the delay. Incoming s light is unaffected by FLC1 and it is reflected by PBS1, the mirrors M1 and

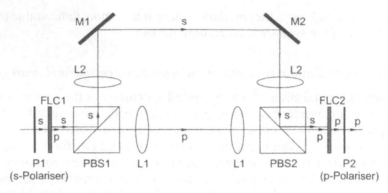

Figure 1.32. Single-bit delay block of a photonic delay line implemented using FLC 90° polarization rotators. Based on the state on FLC1 and FLC2 the light can follow the straight path or the bypass path, introducing a delay.

M2 and finally PBS2, changes into *p* light from FLC2 and is transmitted through P2. FLC2 and P2 act as an active noise filter and substantially reduce polarization leakage effects due to the imperfect 90° polarization rotation by the FLC devices, as explained by Riza and Madamopoulos [89]. The time delay introduced by the bypass of figure 1.32 is theoretically equal to

$$\Delta t = 4 \frac{f_2 - f_1}{c}.$$

Madamopoulos and Riza [89] analysed and demonstrated a three-bit PDL using as the second bit the block of figure 1.32 together with two other delay blocks based on similar principles, but capable of introducing longer and shorter delays to serve as the first and third bits. Bits 1, 2 and 3 were 5.69 ns, 1.67 ns and 8.8 ps, respectively. A switching speed of 35 µs was measured together with optical insertion losses of around 1.5 dB per bit. Later a seven-bit 33 channel PDL was reported [90], based on the same concepts and meeting phased-array antenna requirements for the aerospace industry. Multi-pixel (33-pixel) FLC polarization rotators were employed together with a fibre-optic array combined with graded-index collimators in order to treat simultaneously 33 channels. The system was able to produce 128 different time delays with the least-significant bit accounting for 0.1 ns of delay up to 6.4 ns for the most-significant bit. Optical insertion losses averaged 1.5 dB per bit with low in-channel polarization leakage noise and interchannel crosstalk.

1.7.4.2   *Dynamic arbitrary wavefront generation*

FLC/SLMs can be employed to generate arbitrary wavefronts in a fast reconfigurable way, as shown by Neil *et al* [92, 93]. Using this technique known optical aberrations can be generated dynamically for testing optical systems, and aberrations introduced by the optical system itself can be measured and corrected. To outline the salient points of the method assume that the wavefront $f(x, y) = \exp\{j[\varphi(x, y) + \tau(x, y)]\}$ contains the desired wavefront phase φ and some linear phase tilt τ. Wavefront f is binarized according to the sign of the real part of f, resulting in a binary wavefront g containing the levels ± 1. It has been mathematically proved that g can be expanded in the following Fourier series:

$$g(x, y) = \frac{2}{\pi}\{\exp[j(\varphi + \tau)] + \exp[-j(\varphi + \tau)]$$

$$- \tfrac{1}{3}\exp[j3(\varphi + \tau)] - \tfrac{1}{3}\exp[-j3(\varphi + \tau)] + \cdots\}. \qquad (1.10)$$

The binarized wavefront g corresponding to the analogue wavefront f is displayed on a binary phase FLC/SLM and it is Fourier transformed by a first lens. In the Fourier plane every component present in equation (1.10) will appear in a separate spatial location. Therefore, a suitable spatial filter (pinhole) can be placed to isolate the first positive order, which in turn will be transformed again by a second lens. The second lens is thus producing an analogue wavefront with the desired phase φ and the tilt τ, and the latter can be easily removed. Neil *et al* [92] demonstrated the above method and produced phase screens corresponding to transmission through Kolmogorov turbulence, which are useful in modelling imaging through the atmosphere.

Other advanced applications of the wavefront generation method using FLC/SLMs were proposed for use in confocal microscopy [93]. Wavefront generation can now be used to tune the complex pupil function of the objective lens, modifying the imaging performance of the microscope system and correcting for optical aberrations. Figure 1.33 shows the wavefront generator and its interface to a confocal microscope system. A laser beam is expanded and illuminates a 256×256 LCOS FLC/SLM. Reflected light has a binary phase modulation determined by the displayed pattern and is Fourier transformed by L1. Pinhole PH1 transmits only the first order, which is again Fourier transformed by L2 to provide the desired wavefront. The generated wavefront illuminates the objective lens L3 of the microscope system. This arrangement can correct for aberrations, for instance introduced by focusing deep into thick specimens where spherical aberrations are expected to be predominant, as demonstrated by Neil *et al* [94] in a two-photon microscope. One has to 'pre-aberrate' the objective illumination with a sufficient amount of the conjugate aberration to substantially counteract this effect. Recent applications of the wavefront generation method have been in measuring and correcting for the aberrations encountered when writing three-dimensional bit-oriented photorefractive optical memories in $LiNbO_3$ [96].

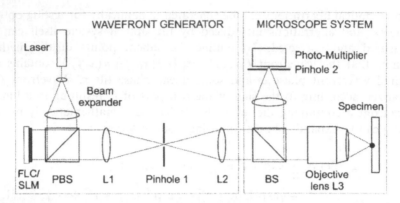

Figure 1.33. The arbitrary wavefront generator and its interface to a microscope system. Wavefront generation is used to tune the pupil function of the objective lens (L3), modifying the imaging performance of the microscope system.

References

[1] Dunmur D A, Fukuda A and Luckhurst G R 2001 *Physical Properties of Liquid Crystals: Nematics* (London: IEE) p 5

[2] Meyer R B, Liebert L, Strzelecki L and Keller P 1975 *J. de Physiques* **36** L69–71

[3] Born M and Wolf E 1999 *Principles of Optics* (Cambridge: Cambridge University Press) p 360, 790

[4] Brown C V and Jones J C 1999 *J. Appl. Phys.* **86** 3333–3341

[5] Mills J T, Gleeson H F, Goodby J W, Hird M, Seed A and Styring P 1998 *J. Mat. Chem.* **8** 2385–2390

[6] Matkin L S, Gleeson H F, Baylis L J, Watson S J, Bowring N, Seed A, Hird M and Goodby J W 2000 *Appl. Phys. Lett.* **77** 340–342

[7] Ulrich D C and Elston S J 1996 *Appl. Phys. Lett.* **68** 185–187

[7a] Guibert L, Keryer G, Mondher Attia A S, MacKenzie H S and de Bougrenet de la Tocnaye J L 1995 *Opt. Eng.* **34** 135–143

[8] de Vries A 1977 *Mol. Cryt. Liq. Cryst.* **41** 27–29

[9] Elston S J 1995 *J. Mod. Opt.* **42** 19–56

[10] Funtschilling J, Stalder M and Schadt M 2000 *Ferroelectrics* **244** 557–564

[11] Wang C H, Kurihara R, Bos P J and Kobayashi S 2001 *J. Appl. Phys.* **90** 4452–4455

[12] Brunet M and Martinot-Lagarde P 1996 *J. Phys. II France* **6** 1687–1725

[13] Glogarova M, Lejcek L, Pavel J and Fousek J 1983 *Mol. Cryst. Liq. Cryst.* **91** 309–325

[14] Brunet M and Williams C 1978 *Ann. Phys.* **3** 237–248

[15] Garoff S and Meyer R B 1977 *Phys. Rev. Lett.* **38** 15 848–851

[16] Clark N A and Lagerwall S T 1980 *Appl. Phys. Lett.* **36** 899–901

[17] Saunders F, C Hughes J R, Pedlingham H A and Towler M J 1989 *Liq. Cryst.* **6** 341–347

[18] Yang K H, Lien A and Chieu T C 1988 *Jap. J. Appl. Phys.* 2022–2025

[19] Ulrich D C 1995 *Domain Formation and Switching in Ferroelectric Liquid Crystals* DPhil Thesis University of Oxford p 14

[20] Srajer G, Pindak R and Patel J S 1991 *Phys. Rev. A* **43** 5744–5747

[21] Surguy P W H, Aycliffe P J, Birch M J, Bone M F, Coulson I, Crossland W A, Hughes J R, Ross P W, Saunders F C and Towler M J 1991 *Ferroelectrics* **122** 63–79

[22] Meyer R B 1977 *Mol. Cryst. Liq. Cryst.* **40** 33–48

[23] Patel J S 1992 *Appl. Phys. Lett.* **60** 3 280–282

[24] Hatano T, Yamamoto K, Takezoe H and Fukuda A 1986 *Jap. J. Appl. Phys.* **25** 1762–1767

[25] Parry-Jones L A and Elston S J 2000 *Phys. Rev. E* **63** art no. 050701(R)

[26] de Gennes P 1968 *Solid State Commun.* **6** 163–165

[27] Nakagawa M 1991 *Jap. J. Appl. Phys.* **30** 1759–1764

[28] Qian T and Taylor P 1999 *Phys. Rev. E* **60** 2978–2984

[29] Parry-Jones L A and Elston S J 2001 *Appl. Phys. Lett.* **79** 2097–2099

[30] Fukuda A 1995 *Proc. Asia Disp. 1995* 61–64

[31] Rudquist P, Lagerwall J P F, Buivydas M, Gouda F, Lagerwall S T, Clark N A, Maclennan J E, Shao R, Coleman D A, Bardon S, Bellini T, Link D R, Natale G, Glaser M A, Walba D M, Wand M D and Chen X H 1999 *J. Mat. Chem.* **9** 1257–1261

[32] Hughes J R and Raynes E P 1993 *Liq. Cryst.* **13** 597–601

[33] Maltese P, Ferrara V and Coccettini A 1995 *Mol. Cryst. Liq. Cryst.* **266** 163–177

[34] Itoh N, Akiyama H, Kawabata Y, Koden M, Miyoshi S, Numao T, Shigeta M, Sugino M, Bradshaw M J, Brown C V, Graham A, Haslam S D, Hughes J R, Jones J C, McDonnell D G, Slaney A J, Bonnett P, Bass P A, Raynes E P and Ulrich D 1998 *Proceeding of the International Display Workshop IDW'98* p 205

[35] Lagerwall S T 1999 *Ferroelectric and Antiferroelectric Liquid Crystals* (Weinheim: Wiley–VCH) pp 383, 390

[36] Efron U (ed) 1994 *Spatial Light Modulator Technology* (New York: Marcel Dekker) p 310

[37] McKnight D J, Johnson K M and Serati R A 1994 *Appl. Optics* **33** 2775–2784

[38] Broomfield S E, Neil M A, Paige E G and Yang G G 1992 *Electr. Lett.* **28** 26–28

[39] Jones R C 1941 *J. Opt. Soc. Am.* **31** 488–503

[40] Fukushima S, Kurokawa T, Matsuo S and Kozawaguchi H 1990 *Optics Lett.* **15** 285–287

[41] Moddel G, Johnson K M, Li W, Rice R A, Pagano-Stauffer L A and Handschy M A 1989 *Appl. Phys. Lett.* **55** 537–539

[42] Kurokawa T and Fukushima S 1992 *Optical and Quantum Electr.* **24** 1151–1163

[43] Clark N A and Lagerwall S T 1991 *Ferroelectric Liquid Crystals: Principles, Properties and Applications* (Philadelphia: Gordon and Breach) p 409

[44] Riza N A and Yuan S 1998 *Electr. Lett.* **34** 1341–1342

[45] Riza N A and Yuan S 1999 *J. Lightwave Technol.* **17** 1575–1584

[46] Berthele P, Fracasso B and de la Tocnaye J L 1998 *Appl. Optics* **37** 5461–5468

[47] Warr S T and Mears R J 1995 *Electr. Lett.* **31** 714–716

[48] Crossland W A, Manolis I G, Redmond M, Tan K L, Wilkinson T D, Holmes M J, Parker T R, Chu H, Croucher J, Handerek V A, Warr S T, Robertson B, Bonas I G, Franklin R, Stace C, White H J, Woolley R A and Henshall G 2000 *J. Lightwave Technol.* **18** 1845–1854

[49] Seldowitz M A, Allebach J P and Sweeney D W 1987 *Appl. Optics* **26** 2788–2798

[50] Mahlab U, Shamir J and Caulfield H J 1991 *Optics Lett.* **16** 648–650

[51] Dames M P, Dowling R J, McKee P and Wood D 1991 *Appl. Optics* **30** 2685–2691

[52] Warr S T and Mears R J 1996 *Ferroelectrics* **181** 53–59

[53] Yamazaki H and Fukushima S 1995 *Appl. Optics* **35** 8137–8143

[54] Tan K L, Crossland W A and Mears R J 2001 *J. Opt. Soc. Am. A* **18** 195–204

[55] Tan K L, Warr S T, Manolis I G, Wilkinson T D, Redmond M M, Crossland W A, Mears R J and Robertson B 2001 *J. Opt. Soc. Am. A* **18** 205–215

[56] Crossland W A and Wilkinson T D 1998 *Handbook of Liquid Crystals: Fundamentals* (Volume 1, chapter 2) ed Demus D, Goodby J, Gray G W, Spiess H W and Vill V (Weinheim: Wiley–VCH) p 763

[57] O'Brien D C, McKnight D J and Fedor A 1996 *Ferroelectrics* **181** 79–86

[58] Patel J S, Saifi M A, Berreman D W, Lin C L, Andreadakis N and Lee S D 1990 *Appl. Phys. Lett.* **57** 1718–1720

[59] Patel J S 1992 *Optics Lett.* **17** 456–458

[60] Liu J Y and Johnson K M 1995 *IEEE Photon. Technol. Lett.* **7** 1309–1311

[61] Sneh A, Johnson K M and Liu J Y 1995 *IEEE Photon. Technol.* **7** 379–381

[62] Sneh A and Johnson K M 1996 *J. Lightwave Technol.* **14** 1067–1080

[63] Choi W K, Davey A B and Crossland W A 1996 *Ferroelectrics* **181** 11–19

[64] Warr S T, Parker M C and Mears R J 1995 *Electr. Lett.* **31** 129–130

[65] Parker M C and Mears R J 1996 *IEEE Photonics Technol. Lett.* **8** 1007–1008

[66] Parker M C, Cohen A D and Mears R J 1998 *J. Lightwave Technol.* **16** 1259–1270

[67] Parker M C, Cohen A D and Mears R J 1997 *IEEE Photonics Technol. Lett.* **9** 529–531

[68] Walker D B, Glytsis E N and Gaylord T K 1996 *Appl. Optics* **35** 3016–3030

[69] Hermann D S, Scalia G, Pitois C, De March F, D'have K, Abbate G, Lindgren M and Hult A 2001 *Opt. Eng.* **40** 2188–2198

[70] Ozaki M, Sadohara Y, Uchiyama Y, Utsumi M and Yoshino K 1993 *Liquid Crystals* **14** 381–387

[71] Clark N A and Handschy M A 1990 *Appl. Phys. Lett.* **57** 1852–1854

[72] D'Alessandro A, Asquini R, Menichella F and Ciminelli C 2001 **372** 353–363

[73] Asquini R and d'Alessandro A 2002 *Mol. Cryst. Liq. Cryst.* **375** 243–251

[74] Fukushima S and Kurokawa T 1991 *IEEE Photonics Tech. Lett.* **3** 682–684

[75] Turner R M, Jared D A, Sharp G D and Johnson K M 1993 *Appl. Optics* **32** 3094–3101

[76] Wilkinson T D, Petillot Y, Mears R J and de Bougrenet de la Tocnaye J L 1995 *Appl. Optics* **34** 1885–1890

[77] Keryer G, de Bougrenet de la Tocnaye J L and Al Falou A 1997 *Appl. Optics* **36** 3043–3055

[78] Iwaki T and Mitsuoka Y 1990 *Optics Lett.* **15** 1218–1220

[80] Petillot Y, Guibert L and de Bougrenet de la Tocnaye J L 1996 *Optics Com.* **126** 213–219

[81] Zhang L, Robinson M G and Johnson K M 1991 *Optics Lett.* **16** 45–47

[82] Gomes C M, Sekine H, Yamazaki T and Kobayashi S 1992 *Neural Networks* **5** 169–177

[83] Wagner K and Slagle T M 1993 *Appl. Optics* **32** 1408–1435

[84] Mao C C and Johnson K M 1993 *Appl. Optics* **32** 1290–1296

[85] Bar-Tana I, Sharpe J P, McKnight D J and Johnson K M 1995 *Optics Lett.* **20** 303–305

[86] Drabik T J and Handschy M A 1990 *Appl. Optics* **29** 5220–5223

[87] Johnson K M, McKnight D J and Underwood I 1993 *IEEE J. Quantum Electron.* **29** 699–714

[88] Mears R J, Crossland W A, Dames M P, Collington J R, Parker M C, Warr S T, Wilkinson T D and Davey A B 1996 *IEEE J. Selected Topics in Quantum Electr.* **2** 35–46

[89] Riza N A and Madamopoulos N 1997 *J. Lightwave Technol.* **15** 1088–1094

[90] Madamopoulos N and Riza N A 1998 *Appl. Optics* **37** 1407–1416

[91] Madamopoulos N and Riza N A 2000 *Appl. Optics* **39** 4168–4181

[92] Neil M A, Booth M J and Wilson T 1998 *Optics Lett.* **23** 1849–1851

[93] Neil M A, Wilson T and Juskaitis R 2000 *J. Microscopy* **197** 219–223

[94] Neil M A, Juskaitis R, Booth M J, Wilson T, Tanaka T and Kawata S 2000 *J. of Microscopy* **200** 105–108

[95] Neil M A, Juskaitis R, Booth M J, Wilson T W, Tanaka T and Kawata S 2002 *Appl. Optics* **41** 1374–1379

Chapter 2

Electro holography and active optics

Nobuyuki Hashimoto

As information devices, such as personal computers and cellular phones among others, have spread far and wide over the past few years, the thin, low-power liquid crystal device (LCD) has become essential for the information age. The fact that the half-wave voltage of liquid crystals is only a few volts, which is far lower than that of solid crystals, makes liquid crystal devices intrinsically good for control of light wavefront, i.e. they can be excellent optical phase modulation devices.

On the other hand, as a result of the latest developments of telecommunications and digital technologies, terabit-class communications and data processing are coming to be a reality. Mass storage three-dimensional display, more specifically the ultimate three-dimensional image processing and video transmission that make use of holography, has come into view of terabit technology. This means that electronic holography has now acquired great importance. Additionally, with the progress and sophistication of technologies that employ light wavefronts, such as optical disk drives and laser printers, active optics, much more than those previously in use, have been brought into use.

This chapter describes electro holography that makes use of liquid crystals for active optics. It also provides in-depth descriptions of the principle of liquid crystal optics that can actively control the light wavefront, and detailed information on its practical applications.

2.1 Electro holography

Holography is the process of recording and reconstructing wavefronts and was devised by Gabor in 1948 [1]. After that, with the invention of lasers, further research was conducted on holography as a three-dimensional display process and optical information processing technique, the spotlight of attention being focused upon the former (three-dimensional display process) as it could provide the ultimate method to form a three-dimensional

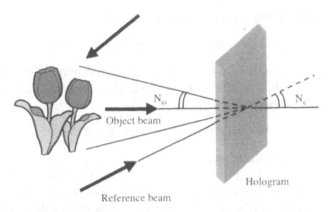

Figure 2.1 Optics for holography.

image of a subject without using the illusion. Holographic recording, however, required a very high resolution, and electronic recording and reconstruction was not possible at that time. Afterwards, holography still had to wait to steal back the spotlight, until 1989 when computerized holovideo appeared on stage [2] and 1991 when a holographic TV using liquid crystal technology was announced [3, 4]. These spurred research into electro holography.

This section describes electro holography that makes use of liquid crystal devices, including the basics of holography and liquid crystals.

2.1.1 The basics of holography [5]

Figure 2.1 shows typical holographic recording optics. A laser beam is divided into two fluxes of light. One of them illuminates an object and its diffracted wave reaches the recording material in the form of an object (signal) beam, and the other is directly irradiated on to the recording material as a reference beam. Interference is produced in this way between the object beam and the reference beam on the recording material. The obtained interference fringe pattern is then recorded and developed, and the resulting image is called a hologram. When this pattern is recorded on a black and white film, it provides an amplitude hologram, and a phase hologram can be produced if the pattern is changed into a transparent phase distribution. With a 90° twisted-nematic liquid crystal device (TN-LCD), an amplitude hologram is produced using a polarizing plate, but to be exact, phase modulation concurrently occurs in this case. With a homogeneous LCD, a pure phase hologram is obtained in principle [6].

Suppose the complex amplitudes to be A and B for the object and reference beams in figure 2.1, respectively, then the complex amplitude distribution T of the interference fringe pattern can be expressed by equation (2.1) in consideration of the superposition of coherent light:

$$T = |A + B|^2 = |A|^2 + |B|^2 + AB^* + A^*B \qquad (2.1)$$

where * denotes complex conjugate. The complex amplitude distribution of the light BT, which is produced by exposing the interference fringe pattern T to the reference beam B, can be shown by

$$BT = (|A|^2 + |B|^2)B + |B|^2A + A^*B^2. \qquad (2.2)$$

Observing the right-hand second term of equation (2.2), we can see that object beam A has been reconstructed. This is the first-order diffracted light. The first term represents the zeroth-order light which is the attenuated component of reference beam B. The third term, the conjugate wave of object beam A after modulation by reference beam B, is consequently the phase-inverted first-order light. It also indicates that equations (2.3) and (2.4) hold for the carrier spatial frequency N_c $(= P^{-1})$ of the interference fringe pattern and the auto-correlation spectral bandwidth $2N_\omega$ of the object, respectively:

$$N_c = \sin \theta_c / \lambda \qquad (2.3)$$

$$N_\omega = \sin \theta_\omega / \lambda \qquad (2.4)$$

where λ is the wavelength of light.

Hence, in terms of information theory, a hologram can be defined as a carrier spatial frequency modulated by the auto-correlation spectrum of an object.

2.1.2   Liquid crystal spatial light modulator for electro holography

For the materialization of electro holography, fine and active control of the light wavefront (phase) is indispensable, and for that purpose a spatial light modulator with high resolution is required. The spatial light modulator is a device that actively modulates complex amplitude of light in a spatial way [7]. In general terms, non-emissive display devices can therefore be said to be spatial light modulators.

This section describes the basics of complex amplitude modulation of light waves by liquid crystal devices and provides a description of a liquid crystal spatial light modulator for electro holography.

2.1.2.1   *Optical characteristics of liquid crystals* [8]

Whilst TN-LCDs in which liquid crystal molecules are twisted are generally used for the purpose of display, homogeneous LCDs with untwisted liquid crystal molecules are usually used for control of the light wavefront as the phase modulator of light waves. With a homogeneous LCD, the phase of

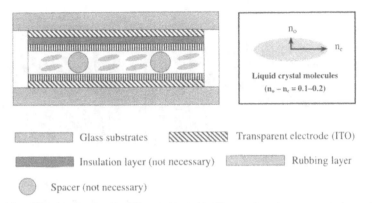

Figure 2.2. Sectional diagram of a typical liquid crystal cell.

light waves only can be modulated without changing the bearing angle of incident linearly polarized light.

Liquid crystal cells.　　Figure 2.2 shows the sectional diagram of typical liquid crystal cells. As can be seen in table 1.1, each film is tens to hundreds nm thick and the layer of these cells is approximately several μm thick, so showing that they are characterized by an optical thin-film structure. Transmittance in the visible region of liquid crystal cells ranges between about 80% and 90%.

Flatness, which is the essential requirement for optical devices, is required as a matter of course for liquid crystal devices when they are used for control of light wavefront. As shown in table 2.1, each individual component of a typical liquid crystal device has a satisfactorily high optical precision. Actually, however, their flatness suffers some deterioration when they are in the process of formation into a cellular structure. This is mainly because of pressurization of cells which is usually performed before and after injection of liquid crystals. Although spacers may diffuse light wavefront, they do not cause any problem because of their limited number (usually, some tens of spacers are used per cm^2).

Table 2.1. Optical properties of a typical liquid crystal cell.

	Refractive index	Thickness
Substrate	1.49–1.52	0.5–1.1 mm
ITO layer	1.7–2	50–200 nm
Insulation layer	1.5–1.7	40–70 nm
Rubbing layer	1.6	40–100 nm
LC layer	1.4–1.8	3–10 μm
Spacer	1.49–1.56	3–10 μm

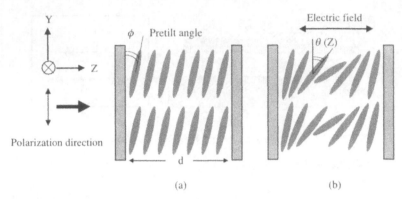

Figure 2.3. Schematic drawing of a homogeneous liquid crystal cell. Without voltage (a) and with voltage (b).

Phase modulation of light [9].   Let us suppose that linearly polarized light is incoming into a set of liquid crystal molecules (see figure 2.3). Here outgoing light is elliptically polarized as it suffers birefringence caused by the refractive anisotropy of these molecules. If, however, the linearly polarized light is incoming parallel to the major axis of these liquid crystal molecules, outgoing light remains linearly polarized as when incoming, without being affected by the birefringence. In this case, the effective refractive factor n_{eff} of liquid crystal molecules can be expressed by equation (2.5), and the incident light propagates as extraordinary rays.

$$n_{\mathrm{eff}} = \frac{n_{\mathrm{e}}^{*} n_{\mathrm{o}}}{(n_{\mathrm{e}}^{2} \sin^{2}(\theta) + n_{\mathrm{o}}^{2} \cos^{2}(\theta))^{1/2}}, \qquad (2.5)$$

where θ is the angle between the major axis of liquid crystal molecules and the axis of polarized light. Hence it is self-evident that n_{eff} is equal to n_{e} if θ is $0°$ and to n_{o} if θ is $90°$.

If linearly polarized light propagates for a distance d across the set of liquid crystal molecules as shown above, the optical path length will be equal to $n_{\mathrm{eff}}d$. Therefore, if n_{eff} is modulated, the phase of an incoming linearly polarized light is modulated. It is also observed from the figure that regardless of the value of θ, n_{eff} always remains equal to n_{o} in relation to linearly polarized light that travels along the direction of the x axis and that the incident light propagates as ordinary rays.

Figure 2.3 shows the structure and principle of operation of a homogeneous LCD. Liquid crystal molecules are contained in the substrates rubbed along the direction of the y axis and are uniformly arranged parallel to this axis (figure 2.3(a)). Here the molecules behave as a continuum. A voltage higher than threshold voltage V_{th} applied across the transparent electrodes along the direction of the z axis causes these molecules to orient their major axes towards the electric field (figure 2.3(b)). If the electric field is sufficiently large, all the molecules are uniformly oriented towards the z axis, that is, they become homeotropic, and the difference in optical path between the optical paths before and after electric field application can be expressed by $(n_e - n_o)d = \Delta nd$, but in point of fact, the molecules present in the proximity of the rubbing interface (several hundred Å) stay static because a strong anchoring force is caused there due to the intermolecular force produced between these molecules and the rubbing film. Therefore, molecular tilt θ becomes a function of Z, being $\theta(Z)$, and its distribution does not depend upon the cell thickness. Optical path length L is then expressed as

$$L = \int n_{\mathrm{eff}}(\theta)\,\mathrm{d}z. \tag{2.6}$$

Nematic liquid crystals show effective value response to an a.c. field and come to a standstill in the same way as when a d.c. field is applied. Usually, an a.c. field is used to drive an LCD because a d.c. field causes impurity ion deviation, and its dielectric anisotropy $\Delta\varepsilon$ slightly depends on the driving frequency. The threshold voltage of a homogeneous LCD is not dependent on its cell thickness but depends upon its dielectric anisotropy $\Delta\varepsilon$, and is around 1 to 2 V in general.

The pretilt angle of liquid crystal molecules affects the voltage threshold (buildup) characteristic of an LCD: the larger the angle, the less steep the voltage buildup curve. The prototype of a single-electrode liquid crystal lens that makes use of this characteristic has already been made.

Complex amplitude modulation of light. Figure 2.4 shows the structure of a 90° twisted-nematic LCD together with the complex amplitude modulation of light. Basically, this LCD is the same as the homogeneous type except that its liquid crystal molecules are rubbed along the direction of the y axis on one substrate and along the direction of the x axis on the other, while their rubbing directions orthogonally cross x and y axes on the respective substrates (see figure 2.4(a)).

Suppose that linearly polarized light is incoming into this LCD in the axial direction. As shown in figure 2.4(a), the incoming linearly polarized light goes propagating while it suffers birefringence at each layer and reiterates the cycle of elliptic and linear polarizations, constantly turning its direction at the same time. Light transmittance T is calculated as below when this LCD is put between two parallel polarizing plates with polarizing axis along the direction of the y axis [10], and without applying voltage to the LCD:

$$T = \sin^2 \frac{0.5\pi(1+\alpha^2)^{1/2}}{(1+\alpha^2)} \qquad (\alpha = 2\Delta nd/\lambda). \tag{2.7}$$

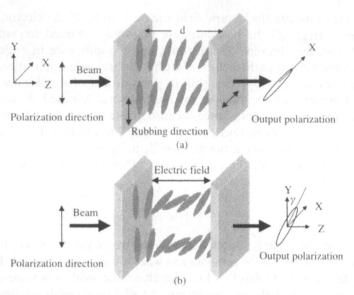

Figure 2.4. Schematic drawing of a 90° twisted-nematic liquid crystal cell. Without voltage (a) and with voltage (b).

Figure 2.5 shows the relation between α and T based on equation (2.7). Light transmittance is zero when equation (2.8) is satisfied:

$$0.5\pi(1 + \alpha^2) = m\pi, \qquad (2.8)$$

where m is an integer. When equation (2.8) is satisfied, outgoing polarized light is linear and polarization occurs along the direction of the x axis. In

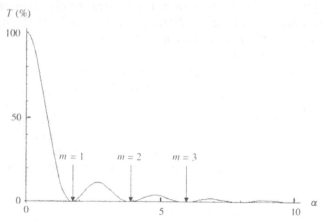

Figure 2.5. Relation between transmittance T and α from equation (2.7).

other words, an ideal 90° optical rotating device with an optical path length of about $n_e d$ can be obtained.

When $m = 1$ holds in equation (2.8) it is called the 'first minimum' state, and when $m = 2$ holds it is called the 'second minimum' state. In the 'first minimum' state, liquid crystals have a better time response but become poorer in the characteristics of visibility angle, wavelength dependency, and temperature.

If a sufficiently large electric field is applied to this LCD along the direction of the z axis, its liquid crystal molecules orient their major axes uniformly towards the z axis, that is, they become homeotropic, losing their optical rotating activity. In this case, outgoing light is linearly polarized along the direction of the y axis, having an optical path length of about $n_o d$. A voltage adjustment allows a middle-point state as shown in figure 2.4(b). After this adjustment, outgoing light is elliptically polarized with a tilt angle of γ from the y axis to the x axis (ellipticity: $\lesssim 10\%$) and its optical path length L also comes to be intermediate ($n_o d < L < n_e d$). When combined with polarizing plates, therefore, it can turn into a complex amplitude modulator that modulates light transmittance and phase simultaneously.

2.1.2.2 Matrix pixel driving [11]

The devices described in the previous section cannot produce the desired patterns spatially because they are not pixellated. In order to generate the desired patterns, devices must be optically addressed [12] or pixellation is required. This section describes the basics of a matrix pixel, which is used in liquid crystal displays and their time-division multiplexing.

Passive matrix driving. Figure 2.6 shows the structure of pixel electrodes used for a passive matrix LCD. The LCD has a matrix of $m \times n$ pixels configured by m pieces of column electrodes and n pieces of row electrodes. Although these column and row electrodes are formed independently on their respective substrates and there is no direct contact with each other, they are electrically connected in the vertical direction through the layer of liquid crystals.

As can be seen from figure 2.6, this electrode structure has the shape of a diffraction grating, which causes diffraction unwelcome to the applications of the hologram device that will be discussed later. Nevertheless, matrixing provides a compensatory great advantage that the number of electrodes can be sharply reduced from $m \times n$ to $m + n$ pieces (otherwise, as many electrode wires as the number of pixels have to be led in).

Time-division multiplexing is generally used to drive a matrix LCD. In time-division multiplexing the selected voltage is applied to row electrodes from the first to the nth row in order of time until the scan of one frame is

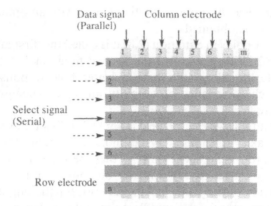

Figure 2.6. Schematic diagram of matrix addressing liquid crystal devices ($m \times n$ pixels).

completed. At the same time, data voltage is applied to all the column electrodes corresponding to one row at one time. Liquid crystal molecules of the pixels whose column and row electrodes have been voltage-applied at the same time give response.

Figure 2.7 shows the driving voltage waveforms focused on one pixel. Because ordinary liquid crystal molecules respond to the applied voltage with effective values, the signal strength for each pixel is controlled by changing the voltage peak value h or voltage application time t_1; the former is called 'pulse height modulation', and the latter is called 'pulse width modulation'. Voltage polarity is inverted each time.

The data voltage applied to the selected row is leaked to other rows while these are not in the selection period. This causes a crosstalk [13]. On the other hand, columns have nothing to do with crosstalk, and therefore the number of columns can be increased without limit in theory.

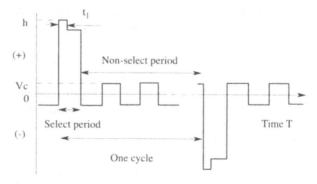

Figure 2.7. Schematic diagram of a driving voltage waveform to a pixel.

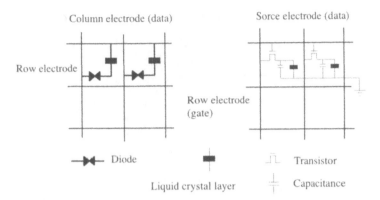

Figure 2.8. Schematic diagram of active-matrix liquid crystal devices using three-terminal devices (a) and two-terminal devices (b).

Active matrix driving. An active matrix LCD is characterized by an independent switch element provided for each pixel. Usually, three-terminal transistors or two-terminal bi-directional diodes are used as switch devices.

In a three-terminal switch device, its transistor source is connected to a data electrode, and its drain is connected to the layer of liquid crystals via a transparent electrode as shown in figure 2.8(a). The opposite transparent electrode is grounded. Therefore, unlike the passive matrix LCD, this opposite transparent electrode is plane and non-pixellated. The driving method of a three-terminal switch device is simple: a selected voltage to turn the gate on is applied to row electrodes while data voltage is applied to data electrodes. During voltage application the gates of non-selected rows remain off, therefore there should be no crosstalk in principle. Actually, however, the problem of crosstalk cannot be ignored because of possible capacity coupling between the transistor and the layer of liquid crystals. In principle, the number of row electrodes can be increased to the level of the transistor on–off ratio (ratio between ON-state resistance and OFF-state resistance: approximately 10^6).

Examples of two-terminal switch devices include diode rings [14, 15] and MIMs (conductor–insulator–conductor) [16]. As shown in figure 2.8(b), each of these devices is connected to the layer of liquid crystals in series, and their pixel electrode is shaped as an x–y grating in the same way as with the passive matrix type. The two-terminal switch device also adopts a simple driving method: a selected voltage higher than the switch device threshold voltage is applied to row electrodes while data voltage is applied to data electrodes. During voltage application, crosstalk is caused by capacity coupling. Although two-terminal devices compare unfavourably with three-terminal types in the driving capacity, they have advantages of

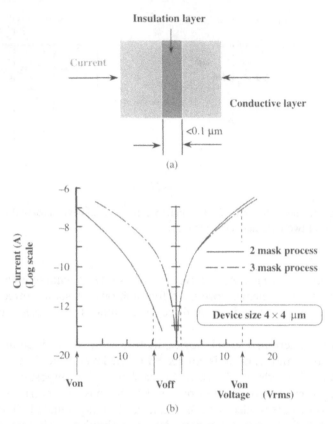

Figure 2.9. A structure of a MIM device (a) and its electrical characteristic (b).

simple structure, suitability for mass production, easy micro-patterning of pixels and easy improvement of fill factor.

2.1.2.3　Liquid crystal spatial light modulator using MIM devices

Characteristics of MIM and its fabrication process [17].　A MIM is one of the most popular thin-film diodes (TFD) featuring nonlinear voltage-to-current characteristics and can be used as a switch device. Figure 2.9 shows its structure and characteristics. An insulating film having a thickness of 0.1 μm or less is held between conductors (see figure 2.9(a)). A thin-film insulator like this provides high insulation at low voltages but when a high voltage is applied, electric charge is excited in accordance with the trap priority in the film, causing its insulation to be reduced. As a result, it works as a voltage switch, as shown in figure 2.9(b). The on–off ratio of a MIM is in the region of 3, which allows around 1000 rows to be driven in theory.

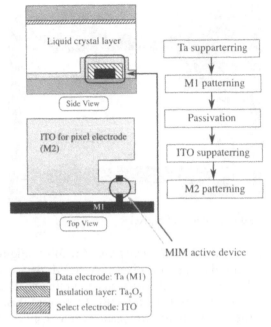

Figure 2.10. Diagram of a MIM fabrication process.

Figure 2.10 shows the fabrication process of a MIM device. Basically, the process consists of film forming by sputtering and photolithographic patterning.

First, a Ta film, which is a conductor, is formed on a glass substrate, followed by patterning of column electrodes and of the film into the lower conductor element (M1 patterning). Next, the upper side of the Ta film is transformed into Ta_2O_5 so that the part changes into an insulator. Then, an ITO, which works as pixel electrodes, is film-formed and patterned (M2 patterning). At this stage a part of the ITO is fine-patterned so that it can be used as the upper conductor element of the MIM, thus featuring a two-mask process.

This simplified process using two masks allows the device surface area to remain unchanged even when some relative dislocation occurs between the two masks (what happens is that the MIM device gets slightly out of position). The process can therefore be used to manufacture microdevices having uniform characteristics in a relatively easy way and is also advantageous for high densification of pixels. But unlike the existing three-mask process, this process requires asymmetrical waves to drive the device as it has asymmetric device characteristics. Figure 2.11 shows a magnified image of one pixel.

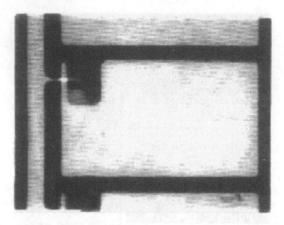

Figure 2.11. Magnified photograph of a pixel.

Characteristics as a spatial light modulator (SLM). Figure 2.12 shows the equivalent circuit diagram of pixels. A MIM device is affected by capacitive crosstalk because of the capacity coupling that occurs between the device and the layer of liquid crystals [18]. To prevent this, it is preferable to provide it with the highest possible capacitive ratio C_R ($=1 + C_L/C_M$). C_R should be as high as possible also in order to allow a sufficient ON-state voltage to be applied to the MIM device during the selection period. Meanwhile, however, an excessively high level of C_R impedes satisfactory electric charging to liquid crystal pixels. In general terms, a C_R value in the region of 4 to 6 is ideal.

A liquid-crystal spatial light modulator (LC-SLM) for electro holography has very fine pixels around 30 µm to 60 µm. Because of this, a level of 2 is the highest possible C_R that can be obtained even if the thickness of its liquid crystal layer is extremely reduced to around 3 µm, and the size of a MIM element to 1 µm square. To solve this, a holding voltage is applied during the non-selection period so that the electric charge injected into

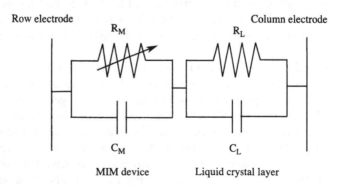

Figure 2.12. Equivalent circuit of each pixel.

Table 2.2. Specifications of a MIM active matrix LCD.

Number of pixels	640×240
Pixel pitch	$30 \times 60\,\mu m$
Cell gap	$4\,\mu m$
Δn	0.1
LC mode	Homogeneous or 90° TN
Input	Composite video signal

liquid crystals can be maintained to minimize possible impact of capacitive crosstalk on the device.

Table 2.2 shows the specifications of an LC-SLM using MIM devices. Video signals convert a spatially modulated pattern (complex amplitude pattern) into a matrix pixel display by video signals. This means this type of LC-SLM works based on the same principle of operation as for a liquid-crystal TV, hence it is called a 'liquid-crystal TV spatial light modulator' (LCTV-SLM) [7].

Figure 2.13 shows the LCTV-SLM using MIM devices, with a photo-montage of the measurement results of flatness (double sensitivity). It shows a flatness of $\lambda/10$ ($\lambda = 633\,nm$) or better.

2.1.3 Electro holography using a liquid crystal TV spatial light modulator [6]

An electro holographic system using LCDs is composed of independent recording and reconstructing optics in which a CCD is used for image

Figure 2.13. LCTV-SLM using MIM active devices.

pickup of interference fringe patterns, and an LCD for hologram display. This section describes the electro holographic system that makes use of a liquid crystal TV spatial light modulator, along with the characteristics of the image reconstructed by this system.

2.1.3.1   Recording optics and reconstructing optics

Recording optics. When an object or its real image is present in the proximity of a recording material, it is called an 'image type hologram', and in this case the image can be reconstructed using white light. Figure 2.14 shows an image-type recording optics and its spectrum. Here it is observed that the real image of an object is performed by a lens near the

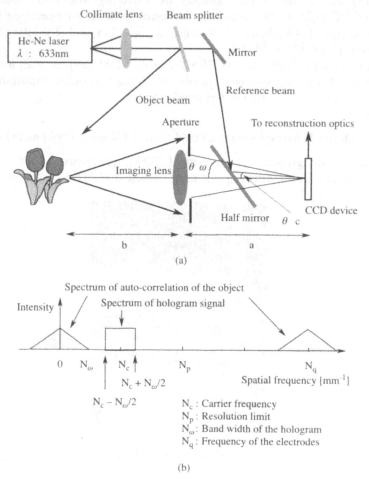

Figure 2.14. Recording optics (a) and its spectrum (b).

CCD. This is called a 'one-step image type' [19] in the terminology of optical hologram.

Figure 2.14(a) shows in a two-dimensional way the concept of how to set spatial frequency for a one-step image-type hologram. A carrier spatial frequency N_c can be controlled by changing the intersecting angle θ_c between an object beam and a reference beam, which can be achieved by adjusting the angle of the half mirror. A spatial frequency bandwidth $2N_\omega$ can be controlled by adjusting the radius r of the aperture stop.

$$N_\omega = \frac{\sin(\tan^{-1}(r/a))}{\lambda} \approx \frac{r}{a\lambda} \tag{2.9}$$

where a is the distance between a lens and a CCD. As it is an image-formation optics, it evidently satisfies the following lens image-formation formula:

$$a^{-1} + b^{-1} = f^{1} \tag{2.10}$$

where b is the distance between a lens and an object.

Given these conditions, the spectral distribution of hologram will be as shown in figure 2.14(b). If it is displayed on a matrix LCD, spectral distribution is cyclically repeated at spatial frequency N_q. Here N_q represents the spatial frequency of the matrix pixel wiring pattern and works as a diffraction grating. Also, because its duty cycle is not 50%, both odd and even high-order spectra are produced. Hence the display of a hologram with a spectral distribution N_h on an LCD brings the resulting spectral distribution $F(N)$ as expressed by equation (2.11).

$$F(N) = N_h * N_q^{-1} \cdot \text{comb}(N/N_q) * \text{sinc}(N \cdot w) \tag{2.11}$$

where $*$ indicates convolution and w is the aperture width of the liquid crystal pixel shown in figure 2.15. Note that in figure 2.14(b) the envelope of 'sinc function' is ignored.

As shown in figure 2.14(b), it is preferable that the auto-correlation spectrum of the object does not overlap spatially with the signal spectrum. In addition, signal bandwidth should not exceed the limit spatial frequency that allows resolution; this is required to avoid the impact of the aliasing

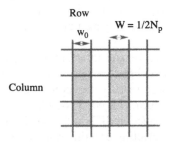

Figure 2.15. Pitch of an electrode (w_0) and aperture of a pixel (w).

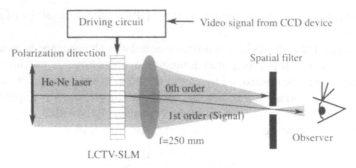

Figure 2.16. Reconstructing optics using a spatial filtering.

version (turn of a signal component that exceeds the maximum spatial frequency allowing display). In the figure, a carrier spatial frequency is set at half of the limit spatial frequency that allows resolution, and a signal bandwidth is set at the largest possible value determined while actually observing reconstructed images. Therefore, it represents the best reconstructed image quality, judged by intuition.

Reconstructing optics. The interference fringe patterns picked up by recording optics are then sent to an LCTV-SLM by video signals. The LCTV-SLM is illuminated by parallel laser beams and an object beam is reconstructed as the first-order diffracted light. At this stage, the spatial frequency for the hologram is limited at a very low level and therefore the zeroth- and first-order light can be difficult to separate in space. In consequence, the reconstructed image is obstructed by the strong zeroth-order light and it is not easy to observe it. It is therefore necessary to remove the zeroth-order light using spatial frequency filtering as described below [3].

Figure 2.16 shows a reconstructing optics that makes use of spatial frequency filtering. As can be seen, a filter is provided at the focal point of the lens. A spatial filter is shown in figure 2.17. An LCD power spectrum by the lens (Fourier spectral intensity) has been stored in this spatial filter. Black dots in the image are diffraction patterns produced by matrix pixel electrodes. These patterns correspond to the zero frequency, N_q (spatial frequency of the matrix pixel wiring pattern), and their higher orders are shown in figure 2.14(b).

After being diffracted by the interference fringe patterns displayed on the LCD, signal beams of light can pass through the spatial filter, so the reconstructed image can be readily observed over the filter without obstruction by the zeroth-order light. If a 90° TN-LCD is used, the zeroth- and first-order lights are each polarized in different ways. In this case, the zeroth-order light can be removed by selecting certain appropriate conditions and using an analyser. This method makes positioning of an optics easier [20].

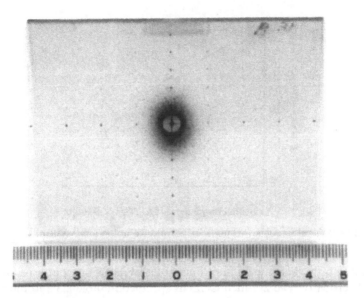

Figure 2.17. Spatial filter.

2.1.3.2   Reconstructed image and its characteristics

The resolution limit of the LCD incorporated in the electro holographic system usually limits its spatial frequency and bandwidth at very low levels. Additionally, its spatial frequency response is affected not only by the LCD but also by the image pickup device and its drivers. Moreover, interference fringe patterns are sampled before display because of matrix pixel configuration.

This section describes the characteristics of a reconstructed image from the electro holographic system, mainly from the viewpoint of information theory.

Reconstructed image.   Spatial frequency response is a critical characteristic of a hologram device, being essential for the design of a holographic optics. Spatial frequency response is measured as described below, using the aforementioned optics.

A sinusoidal grating is projected on the CCD by the two-beam interference method so that it is displayed on the LCTV-SLM. The LCTV-SLM is illuminated by parallel laser beams to produce a diffracted light. The angle of the two beams is regulated to change the spatial frequency of the interference fringe pattern so that the intensity of the first-order diffracted light can be measured in relation to the intensity of the incident light. The resulting spatial frequency response obtained from this measurement is shown in figure 2.18. Here the spatial frequency on the LCD is plotted on the x axis

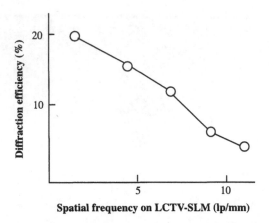

Figure 2.18. Spatial frequency response of the electro holography system.

and diffraction efficiency on the y axis. Interference fringe patterns are represented vertically. The 2/3-inch CCD as specified in table 2.3 and the LCD as specified in table 2.2 were used for this measurement. The spatial frequency response measured by this method represents the frequency response of the whole system, including a CCD and an LCD, and corresponds to the system transfer function.

Next, we discuss the reconstructed image of a three-dimensional object and its characteristics. As this refers to image formation to be used in a recording optics, a biconvex single lens having a focal length of 120 mm is used. The distance between the lens and the CCD is 170 mm, giving an image-formation power of −0.43. Under these conditions, the distance between the object and the CCD is 400 mm. The focal length of the lens for the spatial filter is 250 mm.

The carrier spatial frequency was set on the CCD at about 9 lp/mm horizontally (vertical patterns) and about 5 lp/mm vertically. The effective diameter of the image lens is about 7 mm, which provides a hologram bandwidth (on one side) of about 15 lp/mm on the LCD. This aperture stop diameter was determined on our subjective judgment while actually observing reconstructed images. The intensity ratio between object and

Table 2.3. Specifications of a 2/3 inch CCD device.

Mode	Black and white
Image area	8.8×6.6 mm
Number of pixels	768×490
Pixel pitch	11.4×13.5 μm
Resolution	570×490 TV line
Output	Composite video

reference beams was set at 1:1. We can see from equation (2.2) that the contrast of the interference fringe pattern becomes highest at this intensity ratio.

The results are shown in figure 2.19, where (a) shows the subject used for this image reconstruction, which is about 1 cm square dog-shaped milky

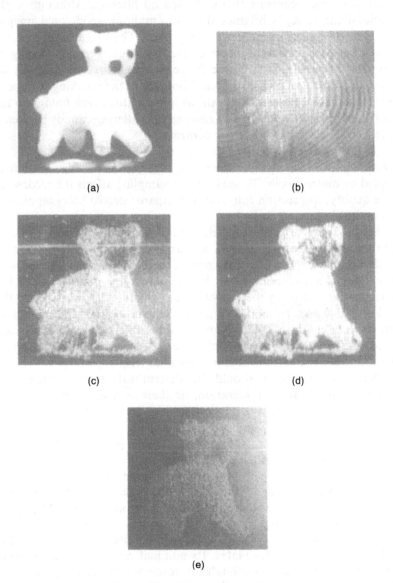

Figure 2.19. Photographs of an object (a), its hologram (b), a reconstructed image from a phase hologram (c), from an amplitude hologram (d) and from an amplitude hologram using white light (e).

glass, and (b) is the hologram interference fringe pattern of that subject, in which the object profile is visible because this is an image-type hologram as mentioned earlier. Figures 2.19(c) and (d) show reconstructed images focusing on phase modulation of light (using a homogeneous LCD) and on amplitude modulation (using a 90° TN-LCD), respectively. These two images show high contrast thanks to spatial filtering. Although a phase-modulated image (c) is brighter than an amplitude-modulated image (d), its quality of contrast is somewhat degraded. The object is put on a step motor and given one turn per second. No image appears during these turns. This is because interference fringe patterns move very rapidly in the same way as with an ordinary optical hologram, thus making it impossible to pick up images. Figure 2.19(e) shows an amplitude-type hologram reconstructed using white light. In this case, spatial filtering cannot be used, and consequently the image has lower contrast.

Comparison with an optical hologram. Interference fringe patterns are sampled by matrix pixels. To see how this sampling affects the reconstructed image quality, this section empirically compares electro holographic images with optical hologram ones.

Spatial frequency values for electro holography are defined as below so that they are ideal for comparison with an optical hologram. First, based on the pixel pitch of an LCD, the maximum theoretical spatial frequencies that allow display are determined to be 17 lp/mm horizontally and 8 lp/mm perpendicularly. Taking the size of LCD, which is 2.2 times larger than a CCD, into account, the above values are converted to 37 lp/mm and 18 lp/mm, respectively, for a CCD. Now, considering a side-band extension due to possible interference by an object beam, carrier spatial frequencies on the LCD are set at half the maximum spatial frequencies.

Next, a bandwidth should be determined, taking carrier spatial frequency values into consideration, as there is a problem with aliasing. Here, however, we take a bandwidth as high as practicable while actually observing experimentally reconstructed images. From these observations result an aperture stop diameter of 6 mm, which provides a bandwidth (on one side) of 28 lp/mm on the CCD. Hence the maximum spatial frequency of the interference fringe pattern becomes 46 lp/mm. This is almost consistent with the resolution limit of the CCD. An image reconstructed from the hologram prepared under these conditions is shown in figure 2.20(a).

The same recording optics described earlier is used as for preparing optical holograms. Agfa 10E75, renowned for its holographic quality, is used for photographic dry plates. Its nominal resolution is approximately 2500 lp/mm, which is high enough for hologram recording. After development, the hologram is immersed in a bleaching solution so as to change it from amplitude-type to phase-type. A reconstructed image is shown in figure 2.20(b).

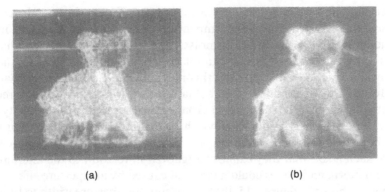

(a)   (b)

Figure 2.20. Reconstructed images from a LCD hologram (a) and an optical hologram (b) using a same recording optics.

Compared with the image reconstructed from an optical hologram, the one from a liquid crystal hologram has slightly higher granularity. This might be attributable to speckles caused by pixel structure, but similar granularity can also be observed in the images reconstructed by white light. Therefore, this high granularity must originate in noise (caused by transmission system, jitters and other noise originating from circuits, and optical noise caused while gradation is being quantized or interference fringe patterns are being sampled). Here optical noises are examined.

First, let us analyse quantization of gradation. As is known, when gradation is quantized it can affect higher-order diffractions and the reconstructed images (resulting in rough surfaces in this case) [21, 22]. To examine this by comparison, figure 2.21 shows two reconstructed images: one reconstructed from 2° gradation display on LCD (a) and the other from 32° gradation display (b). As can be seen from these images, no clear difference is observed in gradation between the two. In addition, as the

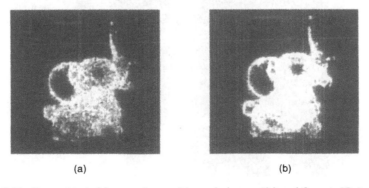

(a)   (b)

Figure 2.21. Reconstructed images from a binary hologram (a) and from a 32 grey-level hologram using homogeneous LCDs with binary or 32 grey-level modulation.

subject has no half tone, there is little impact of quantization. In consequence, there is probably much impact of noises from sampling of interference fringe patterns. Sampling noises may include those due to aliasing and to aperture effect. Aliasing cannot be ignored as the spatial frequency bandwidth was set wider than the theoretical value. But remember that the bandwidth was determined while actually observing experimentally reconstructed images. Also, it is known from experience that image quality can be improved with a higher bandwidth, taking advantage of aliasing to a certain extent. On the other hand, the pixel aperture of LCDs and CCDs are of such a size that they cannot be ignored in relation to the interference fringe pattern, and noises could have been caused by an aperture effect. This means, as shown in figure 2.15, that supposing the aperture width to be w_0, it is obtained by summation of source signal T (spectral distribution of the interference fringe pattern) by giving it delay from 0 to w_0. Hence it can be shown by the following expression, where $\omega = 2\pi N$:

$$T(N)\exp(-j\omega w)\,\mathrm{d}w = T(N)\exp(-j\omega w_0/2)[\exp(j\omega_0/2) - \exp(-j\omega w_0/2)]/j\omega$$

$$= T(N)w_0\exp(-j\omega w_0/2)\operatorname{sinc}(\omega w_0/2). \qquad (2.12)$$

Here one can see from the envelope obtained as a sinc function that a frequency response deteriorates in higher bandwidths. By way of illustration, with a 100% delay ($w_0 = w$), ωw_0 is equal to π at the maximum spatial frequency ($w = \frac{1}{2}N$), thus the response becomes as low as $2/\pi$ as compared with that of a d.c. component.

Figure 2.22 shows a partial enlargement of the interference fringe pattern recorded in an optical hologram. It also shows the equivalent pixel size of CCDs and LCDs. As can be seen from figure 2.22, here pixels are not small enough in relation to the spatial frequency of the interference fringe pattern. Hence, in order to obtain a reconstructed image having a

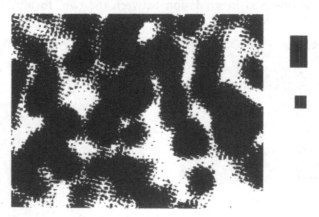

Figure 2.22. Partial magnified photograph of the hologram attached with an equivalent pixel size of a CCD and an LCD.

smoother surface similar to that from the optical hologram, it is essential for the liquid crystal hologram to improve the spatial resolution of both its image pickup and display devices and to increase the number of pixels.

2.2   Active optics

Active optics, capable of active control of light wavefront, will be a key device in future optical electronics. Roughly speaking, either diffraction or refraction effect is used for control of light wavefront: the former has been given a concrete form of holography and diffraction grating, and the latter, kinoform and refractive lens.

    This section describes the basics of diffraction. It also describes the characteristics of active diffractive and refractive optics that make use of liquid crystals, including their applications to optical disk drives, laser scan optics and so forth.

2.2.1   Diffractive optics and refractive optics

2.2.1.1   *The basics of diffraction*

Amplitude rectangular grating.   Figure 2.23 shows an amplitude rectangular grating and its function. As shown in the figure, plane parallel waves enter at an angle of θ in a diffraction grating that is repeating a whole cycle of transparency–opacity at a pitch P (spatial frequency $N = P^{-1}$). In this condition, according to Huygens' principle, spherical waves propagate from each aperture and, at the same time, light waves going out from

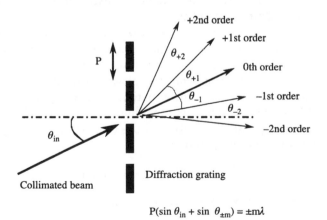

$$P(\sin \theta_{in} + \sin \theta_{\pm m}) = \pm m\lambda$$

Figure 2.23. Theory of diffraction.

adjoining apertures intensify themselves with each other in the direction in which the difference in phase of the two waves is an integer multiple of the wave length. As a result, the incoming light is diffracted towards the direction of $\theta_{\pm m}$ as shown by

$$P(\sin \theta_{\text{in}} + \sin \theta_{\pm m}) = \pm m\lambda \qquad (2.13)$$

where λ is the wavelength and m is an integer. In a diffractive optics, its diffraction efficiency is of vital importance. The mth-order diffraction efficiency is defined as the intensity ratio of the mth diffracted beam to the incoming beam. Generally, the higher the order, the lower the diffraction efficiency. However, design may intensify diffraction of specific order only. The first-order diffraction efficiency of the amplitude rectangular grating (50% duty) is 10.3% in theory.

Although discussion about the complex amplitude of diffracted light in the strict sense of the word requires calculations based on vector diffraction theory [23–25], it can be taken as a question of Fraunhofer diffraction [26] on the basis of scalar theory [27] on the condition that the diffraction grating pitch is equal to or higher than the wavelength and that there is no diffraction grating formed in the direction of the thickness, i.e. if it is not a Bragg diffraction. In short, supposing that plane parallel waves are incoming perpendicularly, the complex amplitude of diffracted light is expressed by Fourier transform of complex amplitude transmittance T of the diffraction grating, and diffraction efficiency is expressed by the square of the absolute value of the obtained amplitude.

Phase rectangular grating [20].   Because of its opaque part, light utilization percentage is very small in the amplitude grating whilst the phase grating provides a large light utilization percentage because it is a transparent grating consisting of a phase distribution alone. Consequently, phase types are normally used for diffractive optics.

Figure 2.24 shows the sectional view of a 50% duty phase rectangular grating with concave–convex distribution shape, where n and n_0 are refraction factors of its substrate and medium, respectively, and x_0 is the grating pitch. Function $f(x)$ that represents the phase distribution of the grating shown in figure 2.24 is expressed by equation (2.14) using 'rect' and 'comb' functions [28] which are defining functions, and its complex amplitude transmittance $T(x)$ (phase distribution in this case) is expressed by equation (2.15) using a convolution operation '$*$':

$$f(x) = \phi_0 + \phi \operatorname{rect}(2x/x_0) * x_0^{-1} \operatorname{comb}(x/x_0) \qquad (2.14)$$

where $\phi_0 = (2\pi/\lambda) \cdot nd$ and $\phi = (2\pi/\lambda) \cdot (n - n_0)d$,

$$T(x) = \exp(if(x)) = \exp(i\phi_0) + (f(x)/\phi) \cdot (\exp(i\phi) - 1) + 1. \qquad (2.15)$$

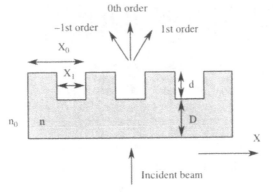

Figure 2.24. Rectangular phase grating.

Because in equation (2.15) exponentials in parentheses have been transformed to constant ϕ exclusively, Fourier transform can be easily calculated and so a complex amplitude distribution $t(\xi)$ of the diffracted light can be expressed by

$$t(\xi) = F[T(x)] = 0.5\,\text{sinc}(\xi/2\xi_0) \sum_m \delta(\xi - m\xi_0) \times \{\exp(i\phi) - 1\} + \delta(\xi)$$

$$(2.16)$$

where $\text{sinc}(x) = \sin(x)/x$, $\xi_0 = 1/x_0$, $\delta(x)$ is the delta function and F is the Fourier transform operator, and m is the integer that shows the order of diffraction.

When m is even in equation (2.16), the value of 'sinc function' becomes zero. Hence diffracted light is produced only in an odd order of diffraction. This is a peculiar phenomenon that occurs only when the duty ratio of a grating is 50%.

The complex amplitude values for the zeroth- and first-order light can be calculated by substituting 0 and 1, respectively, for the integer m ($m = 0$, $m = 1$) in equation (2.16), and can be expressed by equations (2.17) and (2.18), respectively, when the initial phase ϕ_0 is ignored:

$$\text{0th order:} \quad (1/2)(\exp(i\phi) + 1) \qquad (2.17)$$

$$\text{1st order:} \quad (1/\pi)(\exp(i\phi) - 1) \qquad (2.18)$$

where i is a complex unit. As is evident from equations (2.17) and (2.18), both the zeroth- and first-order light are phase-modulated by constant ϕ. The first-order light differs in phase from the zeroth-order by π ($-1 = \exp(i\pi)$). This also applies to the third-order diffraction or higher. Now, focusing on the diffraction efficiency, the efficiency of the first-order light is maximized, being about 41% ($(2/\pi)2 \times 100$), when $\phi = \pi$, that is, phase modulation amount is $\lambda/2$ (λ is the wavelength). These equations also show that the zeroth-order light disappears in this condition. This is known as a Ronchi

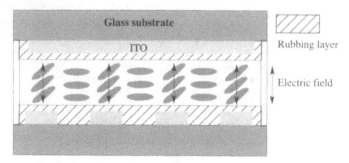

Figure 2.25. Sectional diagram of liquid crystal gratings.

grating in which no zeroth-order light is reconstructed. The diffraction efficiency of the first-order light to ϕ, i.e. $\eta_1(\phi)$, is expressed by

$$\eta_1(\phi) = (2/\pi)^2 \sin^2(\phi/2) \times 100\%. \tag{2.19}$$

The phase rectangular grating can be considered as a close approximation to the saw-teeth grating that can attain the theoretical diffraction efficiency of 100%, which has been embodied by using a two-level binary grating [29]. As its structure is the simplest of all diffractive optics so far available, it can readily provide high spatial frequency devices. Therefore, its use is relatively common where its low light utilization efficiency and higher-order diffraction do not pose any problem to the intended application.

In addition to those mentioned above, sinusoidal amplitude and phase gratings are also commercially available. The maximum diffraction efficiency values of these gratings are 6.3% and 34%, respectively.

2.2.1.2 *Liquid crystal diffractive optics*

Liquid crystal grating. A refractive index distribution is produced in a liquid crystal device by applying an electric field to it via transparent electrodes, so that an optical device that can make electrical control is achieved [30–32]. Therefore, changing of the geometry of these transparent electrodes into that of a diffraction grating materializes a liquid crystal grating [33] which is the basis of a liquid crystal diffractive optics.

Figure 2.25 shows a sectional diagram of a liquid crystal grating which is a one-dimensional grating (50% duty). This is an active diffraction grating that can electrically control phase modulation, i.e. diffraction efficiency.

Table 2.4 outlines the specifications of a liquid crystal grating. Figure 2.26 represents the power spectrum of its diffracted light and figure 2.27 shows the voltage-to-diffraction efficiency chart. It can be seen from figure 2.26 that, as voltage rises, the first-order light intensity increases and the zeroth-order light intensity decreases. Also, when the conditions of a

Table 2.4. Specifications of a liquid crystal grating.

Cell size	20×40 mm
Spatial frequency	5.6 lp/mm
n_e (650 nm)	1.63
n_o (650 nm)	1.49
Cell gap	6.5 µm

Ronchi grating are met, the zeroth-order light disappears. At that stage, its diffraction efficiency is about 37%, which is close to the theoretical value of 41%. If voltage is increased still more to obtain larger phase modulation amounts, the zeroth-order light increases and the first-order light decreases. No power spectrum of diffraction is observed while no voltage application is made to the liquid crystal. Plotted in figure 2.27 are theoretical values, which correspond fairly well to experimental values.

Liquid crystal Fresnel zone plate. A Fresnel zone plate is a diffractive lens popularly known as a Fresnel lens. A Fresnel zone plate is very thin and it

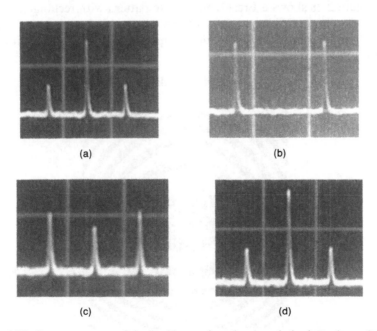

Figure 2.26. Power spectrum of the liquid crystal grating (y: relative intensity, x: 2.3 mm/div). (a) V_{in} 1.75 V_{rms}, phase modulation 0.294π, (b) V_{in} 2.35 V_{rms}, phase modulation π, (c) V_{in} 2.75 V_{rms}, phase modulation 1.253π, (d) V_{in} 3.25 V_{rms}, phase modulation 1.463π.

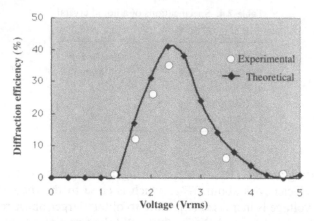

Figure 2.27. Applied voltage versus diffraction efficiency of the liquid crystal gratings.

can easily be arranged in an array. It is also good for photolithography and injection moulding processing, being therefore appropriate for mass production. Because of these characteristics, applications such as a Fourier transformation lens to be used for optical information processing [34] and microlens array [35] have been suggested.

Figure 2.28 shows a Fresnel zone plate pattern with rectangular amplitude transmittance. This pattern works as a lens with focal length f to plane parallel waves having a wavelength λ. Supposing its mth zone radius to be r_M, the following equation holds:

$$r_M = ((M\lambda)^2 + M\lambda f)^{1/2}. \tag{2.20}$$

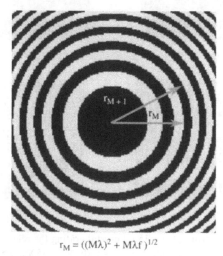

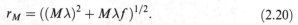

$$r_M = ((M\lambda)^2 + M\lambda f)^{1/2}$$

Figure 2.28. Fresnel zone plate.

Table 2.5. Specifications of a liquid crystal FZP.

Focal length	200 mm (670 nm)
Effective diameter	8 mm
Numerical aperture	0.0041
Zone number	127 line
Minimum line width	16 μm
n_e (670 nm)	1.63
n_o (670 nm)	1.49
Cell gap	3.5 μm

If equation (2.20) is substituted into a phase distribution of 0 and π corresponding to transparency and opacity, it becomes a phase rectangular Fresnel zone plate. The complex amplitude transmittance of a Fresnel zone plate can be considered in the same way as for a diffraction grating.

For both, diffracted light is produced in odd order of diffraction and the diffraction efficiency of the first-order light is about 10.3% for an amplitude type and about 41% for a phase type. Furthermore, their mth-order diffracted beam behaves in a way identical to a lens having a focal length of f/m while their negative-order diffracted beams behave as a concave lens.

Table 2.5 shows the specifications for a phase rectangular liquid crystal Fresnel zone plate that makes use of a transparent electrode pattern as the plate. The electrode pattern of the opposite substrate remains plane and, when voltage is applied, its liquid crystal molecules react to transform it into a Fresnel zone plate. Figure 2.29 shows the electrode pattern. The zone is not perfectly circular but is given a fairly good approximation to a circle by a regular 24-gon. The cross pattern shows a lead electrode wire having a width of 100 μm.

Represented in figure 2.30 is the image-formation spot profile formed by the liquid crystal Fresnel zone plate. Smooth Gaussian spots are observed. Spot intensity depends upon voltage fluctuations but their geometry does not. The photo shows a spot profile, which is almost perfectly consistent with the theoretical value.

Figure 2.31 shows diffraction efficiency in relation to voltage. Diffraction starts at a voltage level of about $1.0 \, V_{rms}$, which is the threshold voltage, reaching the peak (33%) at a level of about $2.6 \, V_{rms}$, and then decreases in turn. The value of $2.6 \, V_{rms}$ satisfies the conditions of a Ronchi grating, thus the diffraction efficiency at that stage should be 41% in theory. The difference between the actual and theoretical values is attributable to the following: a dead space produced by electrode leaders, excessive diffraction caused by the cross pattern and the approximation to a circle by a regular 24-gon.

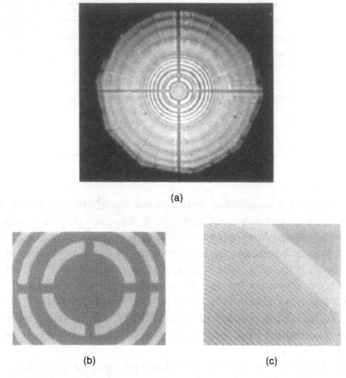

Figure 2.29. ITO electrode patterns of a liquid crystal FZP. (a) A whole image, (b) magnified image (centre) and (c) magnified image (edge).

2.2.1.3 Liquid crystal refractive optics

Refractive optics include existing lenses, prisms and other similar optics, and are based on Snell laws of refraction (see figure 2.32) which refers to a refraction of light that occurs at the boundary surface between media having different refraction factors.

Variable prism. When a potential distribution with linear gradient is applied to a layer of liquid crystals, the layer acts as a prism, permitting light to be refracted with 100% availability. Variation of the distribution gradient allows continuous control of the angle of refraction. Illustrated here is a device approximately provided with a linear potential distribution by connecting low-resistance transparent electrodes with those of high-resistance.

Figure 2.33 shows an electrode pattern for a liquid crystal variable prism. Low-resistance electrodes arranged in the form of a grating are connected with high-resistance electrodes. If therefore a potential difference V is introduced between their ends, a stepped potential distribution as shown

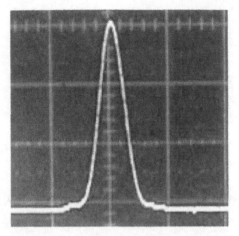

Figure 2.30. Beam spot profile of a liquid crystal FZP (y: relative intensity, x: 40 μm/div).

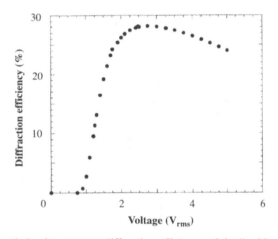

Figure 2.31. Applied voltage versus diffraction efficiency of the liquid crystal FZPs.

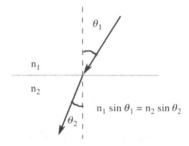

Figure 2.32. Snell laws of refraction.

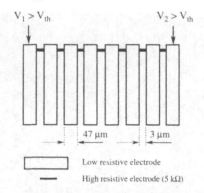

Figure 2.33. Schematic diagram of ITO electrode patterns of a liquid crystal beam deflector.

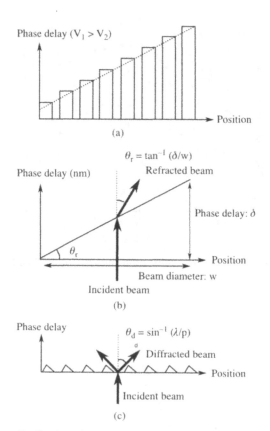

Figure 2.34. Phase distribution of a liquid crystal beam deflector (a) and its components: prism (b) and grating (c).

Table 2.6. Specifications of a liquid crystal variable prism for a beam deflector.

Active area	$5 \times 5\,\mathrm{mm}$
Scan angle	$\theta = 1.1 \times 10^{-3}\,\mathrm{rad}$
n_e (670 nm)	1.7
n_o (670 nm)	1.5
Cell gap	$20\,\mu\mathrm{m}$
Transmittance	88–90% (without AR coating)

in figure 2.34(a) is produced. This distribution can be divided into two: a prismatic potential distribution and a saw-teeth grating potential distribution (see figure 2.34(c)). Hence sufficiently minimized pitch of the grating electrodes allows the saw-teeth grating components to be ignored.

Table 2.6 shows the specifications of the liquid crystal variable prism. One set of electrodes is arranged in the form of a grating and the other set is arranged in a plane pattern. A potential difference V is introduced between their ends, taking the plane pattern as a reference potential (grand level). To obtain a linear potential distribution, a voltage higher than the threshold voltage of the LCD should be applied.

Figure 2.35 shows the relation between potential difference and a refractive angle of light. It is observed from the graph that the liquid crystal variable prism allows linear control of the refractive angle. When the grating pattern space is $5\,\mu\mathrm{m}$, it causes a diffracted beam to increase in quantity and light utilization efficiency to decrease by around 2%. From figure 2.34(b), supposing the phase modulation amount to be δ (nm) and the beam radius to be w, the refractive angle θr can be expressed by $\tan^{-1}(\delta/w)$.

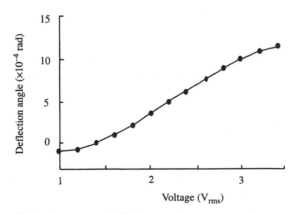

Figure 2.35. Applied voltage versus deflection angle.

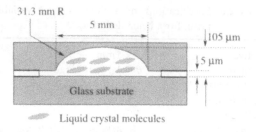

Figure 2.36. Liquid crystal plano-convex lens.

Varifocal lens [36].   A refractive lens with a varifocal function can be obtained by filling liquid crystals in a lens-shaped clear plate. Figure 2.36 illustrates a plano-convex lens. The specifications for this lens are outlined in table 2.7, and its voltage-to-focal length chart is shown in figure 2.37. As it is not easy to increase the difference in refraction factor between liquid crystal and substrates, a lens of this type having a large numerical aperture (NA) can be difficult to obtain. In addition, rubbing of liquid crystals is also difficult because there is a large difference in thickness between their cell centre and periphery. In the case of a homogenous LCD, its threshold voltage does not depend upon liquid crystal cell thickness. Also a liquid crystal tilt distribution is proportional to an external voltage. As a result, a phase distribution similar to the convex shape can be obtained using plane transparent electrodes on both the upper and lower substrates. Figure 2.38 shows a device with a polarizing plate attached. We can see the magnified images through an active area of the LCD. Figure 2.39 shows the image-formation spot profile formed by the device. The image shows that the spot profile is perfectly consistent with the theoretical value and that it is geometrically aplanatic.

2.2.2   Application to optical pickup

Optical pickup is a diffraction-limited optics, which is geometrically aplanatic. Therefore, wave-optic aberration, i.e. wave aberration, must be taken into consideration. Most of factors that cause aberration are attributable to optical

Table 2.7. Specifications of a liquid crystal varifocal lens.

Effective diameter	5 mm
Focal length	150 mm–8 (variable)
Cell gap	5 μm (min) 105 μm (max)
n_e (670 nm)	1.7
n_o (670 nm)	1.5

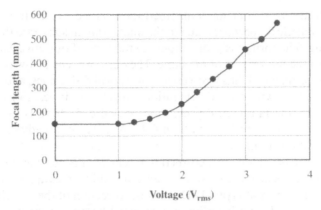

Figure 2.37. Applied voltage versus focal length of liquid crystal lens.

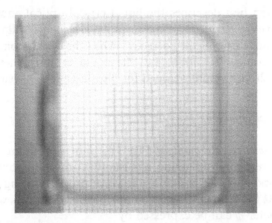

Figure 2.38. Magnified image by liquid crystal (plano-convex) lens.

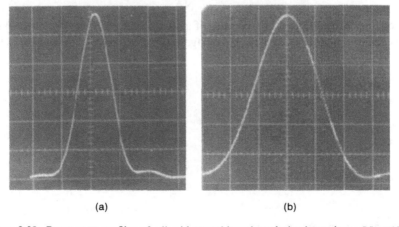

(a)   (b)

Figure 2.39. Beam spot profiles of a liquid crystal lens (y: relative intensity x: 25 μm/div).

disk substrates. In particular, coma aberration caused by substrate warp and/ or tilt and spherical aberration due to the fluctuation in substrate thickness are outstanding. The former is proportional to the cube of objective lens NA, and the latter to the biquadrate of the same. The objective lens NA was about 0.45 for a CD system, which has first-generation optical disk drives, and about 0.65 for DVD system, with second-generation optical disk drives. For a blue-laser DVD system, the latest third-generation optical disk drive is used, with an objective lens having an NA of about 0.85. As a result, an adverse impact of coma aberration upon the RF signal poses a problem to DVD systems. To solve this problem, a disk tilt correcting mechanism is coming into use. This mechanism corrects the tilt by mechanically tilting pickups or disk tables using motors. For multilayer DVD systems, correction of the spherical aberration that is caused by a layer jump is essential because it must read various different layers.

This section first describes the liquid crystal optical equalizer, which is used to improve the frequency response of DVD systems, followed by descriptions of the liquid crystal tilt corrector, which has been commercialized for correction of coma aberration, and the liquid crystal spherical aberration corrector, which has been developed for blue-laser DVD systems.

2.2.2.1   Optical equalizer

The basics of an optical equalizer [37].   As shown in figure 2.40, an optics designed geometrically aplanatic apparently forms infinitesimal spots but in fact these spots are finite due to the wave property of light. In this case, supposing that the incident light is monochromatic and its wavelength is λ, the spot diameter ϕ can be expressed by

$$\phi = \frac{k\lambda}{\text{NA}} \qquad (2.21)$$

where NA is the numerical aperture of a lens which is defined by $n\sin(\theta)$ and n is the refraction factor of the image side medium of the optics, which

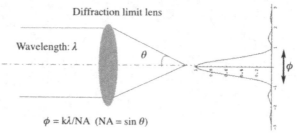

Figure 2.40. Diffraction limit of optics.

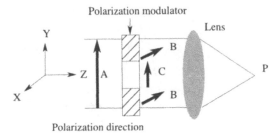

Figure 2.41. Polarization modulator.

usually is 1 (in air). Constant k is inherent in optics and is dependent upon the incoming light intensity distribution. By way of illustrative example, this constant is 1.22 for plane waves of homogeneous intensity distribution.

As can be seen from equation (2.21), the resolution limit of an optics is uniquely identified once the light wavelength and the NA of its lens are defined, and is also called the diffraction limit. On the other hand, the shorter the wavelength (the larger the wavenumber vector), the higher the resolution of the optics. Super-resolution, which means to obtain optical microspots over the diffraction limit, can be achieved by increasing the wavenumber vector. In practice, this is achieved by using light having a shorter wavelength or filling the image side space with a medium having a high refraction factor so as to increase the number of waves. Optical microspots finer than those of the diffraction limit can be obtained in the main-lobe by installing an opaque filter at the optics pupil centre. As light passing through the pupil centre is composed of low spatial frequency components, filtering produces an equalizer effect that intensifies higher frequencies. This method, however, has a disadvantage that it causes strong side-lobes.

The effect of the optical equalizer that makes use of polarization is described below using figures 2.41 and 2.42. Linearly polarized light A is incoming along the direction of the y axis, passes through the polarization modulator and is transformed into a group of linearly polarized light (figure 2.41). Let us suppose that, at the same time, linearly polarized light B passes through the periphery of the modulator and then turns towards the x axis whilst linearly polarized light C passes through the centre of the modulator and then goes ahead along the direction of the y axis without turning. These linearly polarized beams of light are converged by the lens so that optical spots are formed at the point P.

Figure 2.42(a) shows optical spots formed by the respective polarized light of A, B and C, represented by amplitude conditions. An optical spot by A corresponds to a diffraction limit spot in the same way as shown in figure 2.40. An optical spot by B has strong side-lobes in the same way as when the pupil centre is screened with an opaque filter. Here optical spots by B and C do not interfere with each other because light B and C are at right angles to each other. For this reason, their amplitude values do not

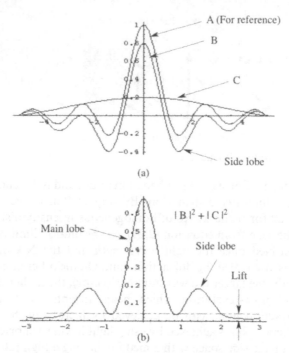

(a)

(b)

Figure 2.42. Beam spot profiles in amplitude (a) and in intensity (b) using a polarization modulator.

add up but intensity values do. In consequence, the optical spot profile as shown in figure 2.42(b) is formed at point *P*.

Now let us discuss the polarized state of main-lobes and side-lobes using figure 2.42(a). Both of them are composites of linearly polarized beams *B* and *C*. Focusing on side-lobes, their amplitude value is negative and polarization vector is −*B*. In this case, if phase difference between *B* and *C* is an integer multiple of π, the resultant force also becomes a linearly polarized light, and the resulting polarized state is as shown in figure 2.43. Hence a polarizing filter can be used to regulate the percentages of both these lobes.

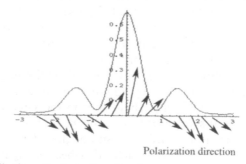

Polarization direction

Figure 2.43. Distribution of polarization directions.

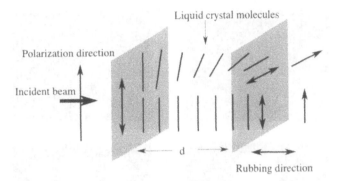

Figure 2.44. Hybrid (90° TN-homogeneous) aligned liquid crystal cell.

Optical equalizer using LCD. The polarization modulator mentioned earlier can be obtained using a hybrid rubbing liquid crystal device as shown in figure 2.44. It is a hybrid of a 90° TN-LCD and a homogeneous LCD. When linearly polarized light is incoming in this device in parallel with the rubbing axis, it is transformed into a pair of linearly polarized lights of the same phase, perpendicular to each other, as the optical path is identically $n_e d$ in both regions.

Figure 2.45 shows some examples of converging spot profiles recorded by the optics shown in figure 2.41 using the hybrid rubbing LCD. Table 2.8 describes the specifications of the LCD. The 90° TN-LCD region satisfies the second minimum conditions around the wavelength of 670 nm and so outgoing light is linearly polarized almost perfectly. The spot profiles in the photo were generated using parallel beams of light (4 mm beam diameter and 670 nm wavelength) as incoming light and converging them with a lens of focal length of 300 mm after their passage through the LCD.

Figure 2.45(a) shows the spot profile without the LCD, corresponding to diffraction limit spots; figure 2.45(b) shows the spot profile recorded using a shielding mask having a geometry identical to the stripe zone (homogeneous LCD region) and represents the conventional super-resolution. Figure 2.45(c) shows the spot profile with the LCD; figure 2.45(d) is the profile obtained from the same device as for (c) plus polarizing plates for compression of sidelobes. Microspots over the diffraction limit are observed. Figure 2.45(e) is the profile after compression of the main-lobe.

This method cannot remove side-lobes perfectly. Furthermore, compression of side-lobes requires a thicker beam diameter. This is because both the main-lobe and the side-lobe have polarization distribution and so they cannot be controlled independently from each other to perfection.

Liquid crystal optical equalizer for optical pickup [38]. This section describes the results of application of the liquid crystal optical equalizer to optical pickups. What is evaluated here is the RF signal. The RF signal is a signal

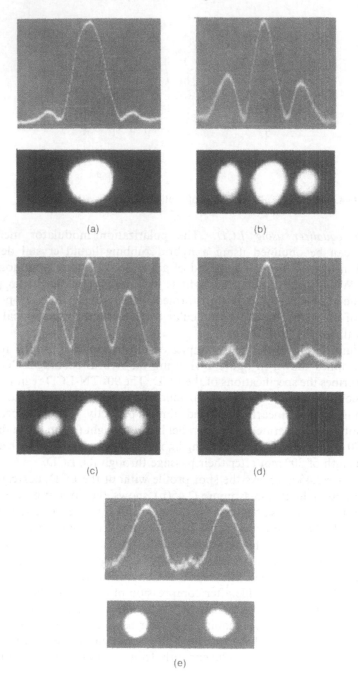

Figure 2.45. Beam spot profiles: Diffraction limit (a), conventional super-resolution (b), liquid crystal optical equalizer (c), suppressed side-lobe (d), and suppressed main-lobe (e) (y: relative intensity, x: 25 μm/div).

Table 2.8. Specifications of a liquid crystal optical equalizer.

Substrate thickness	0.5 mm
Cell size	9 × 6 mm
Effective area	5 × 5 (4f) mm
Δn	0.12
Cell gap	10.5 μm
Homogeneous area	1 mm (stripe)

beam transformed into an electrical signal after being diffracted in optical disk pits. The specifications for the LCD used for this purpose are shown in table 2.8. Figure 2.46 shows the optics for DVD test systems. Parallel laser beams (670 nm wavelength) pass through the LCD and beam splitter 1, and are converged upon an optical disk through an NA of 0.55 objective lens, and then reflected beams of light go back through the objective lens after being diffracted in optical disk pits. Their optical path is split at beam splitters 1 and 2 and the beams are converged upon optical sensors 1 and 2. Sensor 1 detects servo signals to feed them back to the objective lens actuator whilst sensor 2 detects RF signals via its analyser.

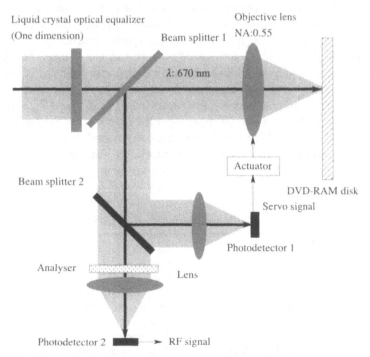

Figure 2.46. Schematic diagram of a DVD test system for a liquid crystal optical equalizer.

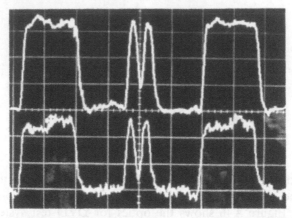

Figure 2.47. RF signals (upper: without equalizer, lower: with equalizer) (courtesy of T D Milster, University of Arizona).

Figure 2.47 shows RF signal waveforms. We can see that 1.5 MHz signal contrast has improved. The optical equalizer in use is a one-dimensional device parallel to the tracks. The relative bearing between the analyser and the linearly polarized beam incoming into the liquid crystal device is 35°.

Next, we discuss the simulation results of a system transfer function when the liquid crystal optical equalizer is used. The conditions we used for this simulation are given in table 2.9, and the results obtained from the simulation in figure 2.48, in which the pit spatial frequency is plotted in the x axis and normalized RF signal intensity in the y axis. The bearing angle of the analyser is relative to the polarization bearing of incoming beams. Hence 90° bearing corresponds to read-out by the shielding-type optical equalizer (conventional super-resolution), and the 45° to read-out due to diffraction limit. The 20° bearing is for read-out by the liquid crystal optical equalizer for which side-lobes were compressed; here we can see that higher frequencies are relatively intensified, that is, the equalizer has worked successfully as intended and its waveforms can be regulated by varying the bearing angle of the analyser. An electrical equalizer is generally used for

Table 2.9. Simulation conditions of a system transfer function.

NA of objectives	0.6
Beam profile	Gaussian
Wave length	650 nm
Pit pitch	4.33 µm (max) 0.54 µm (min)
Pit length	2.6 µm (max) 0.33 µm (min)
Pit width	0.4 µm
Pit mode	Amplitude
Direction of equalizer	Parallel to the track

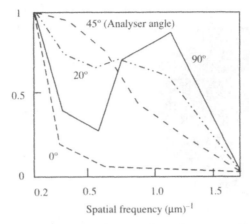

Figure 2.48. System transfer function of a optical equalizer system (simulation) (courtesy of Professor T Milster, University of Arizona).

correction of the MTF response of DVD sytems but it is unfavourable as it affects not only the signal amplitude but also its phase. The optical equalizer makes a favourable comparison as it does not affect signal phase characteristic. But it has nothing to do with spatial frequency cut-off, so that resolution limit remains unchanged.

2.2.2.2 Coma aberration corrector

Correction of coma aberration caused by disk tilt [39]. Optical disks counts on pit information on the back of their substrates are shown in figure 2.49. For this reason, the objective lens of an optics is designed aplanatic, including its plane parallel plates which usually have a specific thickness according to the application (0.6 mm for DVD systems; refraction factor of 1.49). Because of this, they are susceptible to coma aberration when the disk substrate tilts. Supposing the third-order coma aberration factor to be A, wave aberration $W(r, \phi)$ can be expressed by Zernike's polynomial as

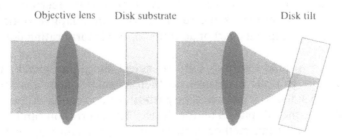

Figure 2.49. Schematic diagram of aberration caused by disk tilt in an optical disk system.

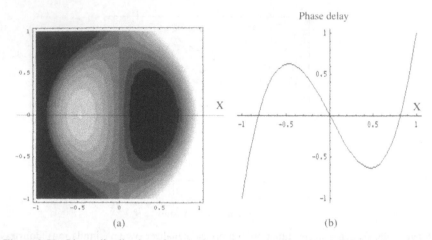

Figure 2.50. Phase distribution of coma aberration: two-dimensional distribution (a) x cut profile (b).

below [40]:

$$W(r,\phi) = A(3r^3 - 2r)\cos\phi$$

$$A = \frac{-t(n^2 - 1)n^2 \sin\beta\cos\beta \cdot NA^3}{6(n^2 - \sin^2\beta)^{5/2}} \tag{2.23}$$

where β is the tilt angle of a substrate, t is the thickness of a substrate, and NA is the numerical aperture of an objective lens.

Figure 2.50 shows the phase distribution of wave-optic coma aberration on the entrance pupil that can be calculated using equation (2.23). Figures 2.50(a) and (b) show the two-dimensional distribution chart and the sectional view of its x axis, respectively. By placing a phase plate at the entrance pupil to correct this phase distribution, coma aberration can be compensated.

Correction of coma aberration by active liquid crystal optics. The strong dielectric anisotropy, one of the characteristics of liquid crystal molecules, can be used for control of light wavefront via an external electric field. In other words, liquid crystals can actively correct wave aberration of optics and are very promising as the future device for adaptive optics. Figure 2.51 shows a conceptual chart of adaptive optics incorporating liquid crystal devices [41].

A beam profile sensor is used to detect the point spread functional produced due to optic aberration to then calculate phase distribution on the liquid crystal device surface. The pattern for correction of the calculated distribution is then displayed on the device. The aberration can be corrected in this way. Whilst this method is more advantageous than the exiting active mirror method [42] in that no mechanical drive is required, it involves

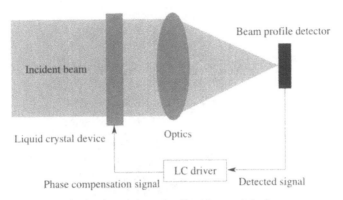

Figure 2.51. Concept of adaptive optics using liquid crystal devices.

problems awaiting solution, such as control available for linearly polarized beams only and lowering of light utilization percentage.

Use of matrix-type liquid crystal devices allows correction of any type of aberration. If the requirement is to correct coma aberration or other specific aberration exclusively, a segmented electrode pattern specially designed for such intended use can be used, so there is no problem. By way of illustration, described below is the liquid crystal tilt correction device which counts on electrode patterns to approximate the phase distribution shown in figure 2.50.

Figure 2.52 illustrates the liquid crystal tilt corrector, and figure 2.53 shows the transparent electrode patterns incorporated in the device. As can be seen from the figure, transparent electrodes on the substrate on one side are segmented into three while those on the opposite side remain plane. The area of these plane electrodes having the same potential is shown with the same numbers in the figure. As a result, some residual aberration remains even after correction is performed (see figures 2.53(b)

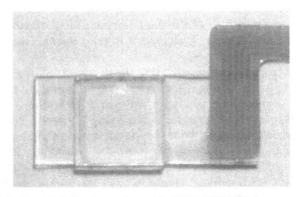

Figure 2.52. Liquid crystal tilt corrector used in DVD-ROM drives.

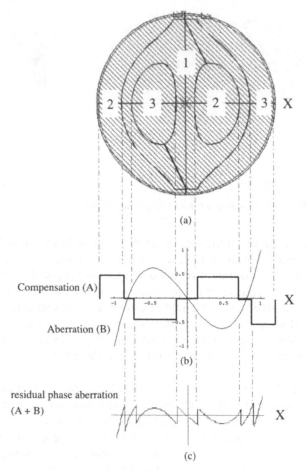

Figure 2.53. ITO patterns of a liquid crystal tilt corrector (a), phase delay profile in the *x* direction (b) and residual phase aberration (c).

and (c)). The specifications of this liquid crystal tilt corrector are given in table 2.10. Its response time is equivalent to the time necessary to obtain a variation of 100 nm in phase.

Phase modulation characteristics of the device are shown in figure 2.54. From the figure we can see that sufficiently large amounts of phase modulation, 650 nm or above, are produced between −20 °C and +80 °C. On the other hand, owing to the limitation of the laser power of optical pickups, the light transmittance of the liquid crystal device is of vital importance. In addition, a liquid crystal device is normally subject to fluctuations of transmittance due to its thin-film structure and variation in its effective refraction factor while driving. Hence its light transmittance characteristics are to be taken into consideration when deciding the thickness of each film

Table 2.10. Specifications of a liquid crystal tilt corrector.

Substrate thickness	0.5 mm
LC mode	Homogeneous
Δn	0.205
Cell gap	4.5 µm
Transmittance	88–90% (650–780 nm) without AR coating
Flatness	$\lambda/10$
Operating temp.	−30 °C to 85 °C
Response time	Under 400 ms (at −20 °C)

and refraction factor. Figure 2.55 shows an example of light reflection characteristics of a liquid crystal device obtained by simulation, and the simulation conditions are given in table 2.11. It can be seen from figure 2.55 that the reflection factor reaches its saddle point when the ITO film used as transparent electrodes is approximately 180 mm thick. Figure 2.56 shows the measured light transmittance values of the actual liquid crystal tilt corrector in discussion. As a result of the optimal design of thin films, the variation in its light transmittance caused by drive load has been controlled down to around 1.5%.

Figure 2.57 illustrates the results of evaluation of the effect of the liquid crystal tilt corrector in question, conducted using RF signals (eye-pattern). Tilt correction was performed perpendicularly to the disk tracks (radial tilt correction). Here the disk tilt angle is 1°. It is observed that RF signal waveforms are improved while this tilt corrector is working.

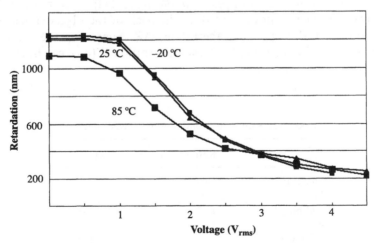

Figure 2.54. Applied voltage versus retardation of the liquid crystal tilt corrector ($\Delta n = 0.205$, $t = 4.5$ µm).

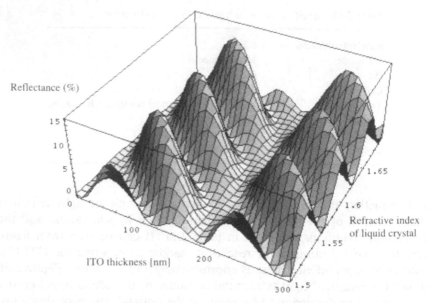

Figure 2.55. Reflectance of liquid crystal devices as an active thin-film device.

Figure 2.58 is a representation of the optical pickup for a car-mounted DVD system already on the market, in which this device has been incorporated.

2.2.2.3 *Spherical aberration corrector* [43]

Spherical aberration caused by plane parallel plates [44].   As shown in figure 2.59, plane parallel plates installed on the back of the objective lens cause defocusing and also spherical aberration, ΔZ. Supposing the third-order spherical aberration factor to be B, wave aberration $W(r, \phi)$ can be expressed by Zernike's polynomial as

$$W(r, \phi) = B(6r^4 - 6r^2) + 1, \qquad B = t(n^2 - 1)\mathrm{NA}^4/8n^3 \qquad (2.24)$$

Table 2.11. Simulation condition of transmittance.

	Refractive index	Thickness
Substrate	1.49	
ITO	1.9	
Rubbing layer	1.63	700 nm
n_e	1.72	
LC layer		5 μm

(Pre-tilt angle 5°)

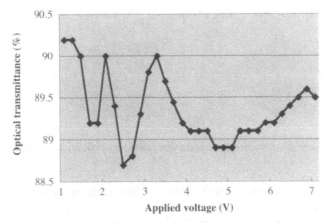

Figure 2.56. Applied voltage versus optical transmittance of the liquid crystal tilt corrector.

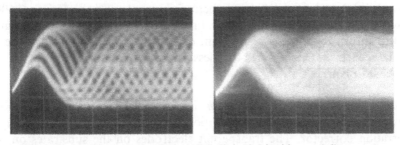

Figure 2.57. RF signal (eye patterns) when disk tilt is 1°. Liquid crystal tilt corrector on (a) and off (b) (courtesy of Pioneer Components Division).

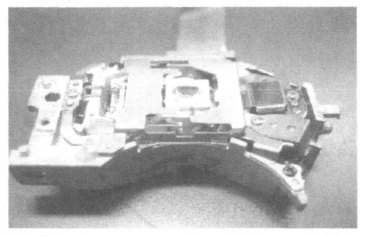

Figure 2.58. A DVD-ROM pickup using a liquid crystal tilt corrector (courtesy of Pioneer Component Division).

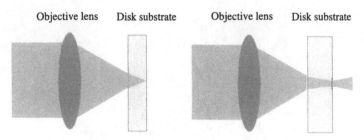

Figure 2.59. Schematic diagram of aberration caused by disk thickness variation.

where t is the thickness of a substrate and NA is the numerical apertures of the objective lens.

Figure 2.60 shows an example of wave aberration, which occurs when the NA of the objective lens is 0.85, the substrate is 100 μm thick and its refraction factor is 1.5. In the figure, coordinates on entrance pupil are plotted on the x axis and wave aberration on the y axis. Figure 2.60(a) shows the values on a Gauss plane (a theoretical focal plane), and figure 2.60(b) shows those on a plane 3 μm defocused from the Gauss plane. It can be seen from these charts that less phase correction is required on the defocused plane.

Liquid crystal spherical aberration corrector. Figure 2.61 shows the transparent electrode patterns incorporated in the liquid crystal spherical aberration corrector. The transparent electrodes on the substrates on one side are left plane, so that they can be patterned for correction of coma aberration, if desired, and transform the device into a hybrid type.

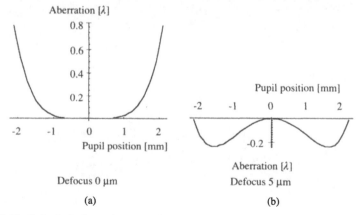

Figure 2.60. Spherical aberration caused by an optical flat: Gauss plane (a) and best plan (b). NA of objective lens: 0.85, thickness = 100 μm ($n = 1.5$).

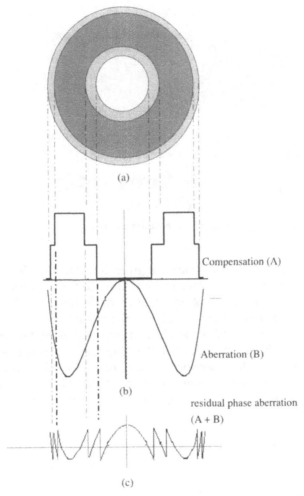

Figure 2.61. ITO pattern of a liquid crystal spherical corrector (a), phase delay profile in the *x* direction (b) and residual phase aberration (c).

Table 2.12. Specifications of a liquid crystal spherical aberration corrector.

Substrates	0.5 mm (thickness)
LC mode	Homogeneous
Δn	0.22
Cell gap	4.0 µm
Transmittance	84–88% (405 nm) without AR coating
Flatness	$\lambda/10$
Operating temp.	−30 °C to 85 °C
Response time	Under 400 ms (−20 °C)

Figure 2.62. Liquid crystal spherical aberration corrector.

The specifications for this device are given in table 2.12. Wave aberration produced at the defocused plane is approximated for correction. The results of the correction are shown in figure 2.61, which illustrates the actual wavefront and its approximation (a, b) and the residual wave aberration after correction (c). Figure 2.62 is the photographic representation of the device. The device has been developed for the blue laser beam whose wavelength is 405 nm. As the light absorption coefficient of ITO films, commonly used as transparent electrodes, increases several times when wavelength falls to 405 nm, its transmittance is reduced, which is the problem to be solved.

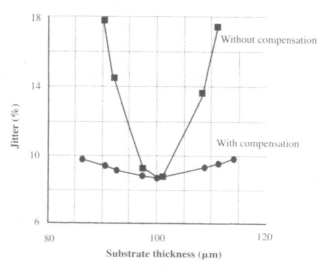

Figure 2.63. Substrate thickness variation versus jitter.

Figure 2.63 shows the signal-regenerating characteristic of an optical pickup in which this device has been incorporated. This is a jitter characteristic obtained after a variation of 30 μm in disk substrate thickness, using a laser beam with a wavelength of 405 nm and an NA of 0.85 objective lens. It can be seen that the characteristic is improved in good measure while this liquid crystal spherical aberration corrector device is working.

2.3 Conclusion

Liquid crystal display devices have made a marked advance and have already won a big market. On the other hand, mass production of organic electroluminescent and plasma display devices is on the move, and electric emissive displays and various other new types of device are showing up on the market.

Nevertheless, in the control of light wavefront, liquid crystal devices outshine all these newcomers because these new devices cannot control light wavefront, in theory. Only an LCD intrinsically holds an unchallenged position in this field.

Furthermore, although we could not discuss this here, liquid crystal switches and filters for optical communications are already arriving on the market. Also, their application to optical face recognition systems is already taking shape and is expected to be commercialized soon [45]. In addition, in the stellar coronagraph, which is a system to eliminate star noises in planetary observation activities, polarization modulation filters that make use of ferroelectric liquid crystals are attracting a good deal of attention [46].

In terms of active optics, the most powerful and toughest rival of LCDs will certainly be an optical MEMS [47]. An optical MEMS surpasses an LCD in temperature characteristics and response speed whilst an LCD outmatches an optical MEMS in the aspects of reliability, costs and mass production.

Finally, we hope that research and development activities for performance upgrading and exploitation of new application fields, making the most of the advantages of an LCD, will be conducted still more actively so that an LCD continues growing as one of key technologies in the field of optoelectronics.

References

[1] Gabor D 1948 *Nature* **161** 177
[2] Kollin J S, Benton S A and Jepsen M L 1989 *Proc. of the 2nd International Congress of Optical Sciences and Engineering*

116   *Electro holography and active optics*

[3] Hashimoto N, Morokawa S and Kitamura K 1991 *Proc. SPIE* **1461** 291–302
[4] Sato K, Higuchi H and Katsuma H 1992 *Proc. SPIE* **1667** 19–31
[5] Hariharan P 1984 *Optical Holography* (Cambridge: Cambridge University Press)
[6] Hashimoto N and Morokawa S 1993 *J. Electronic Imaging* **2** 93–99
[7] Liu H K, Davis J A and Lily R A 1985 *Opt. Lett.* **10** 635–637
[8] Yeh P and Gu C 1999 *Optics of Liquid Crystal Displays* (New York: Wiley)
[9] Klaus W, Hashimoto N, Kodate K and Kamiya T 1994 *Opt. Rev.* **1** 113–117
[10] Gooch C H and Tarry H A 1975 *J. Phys. D: Appl. Phys.* **8** 1575–1584
[11] Dargent B *et al* 1977 *SID Digest* (SID77) 60–61
[12] Belha W H *et al* 1975 *Opt. Eng.* **14** 371–384
[13] Inukai T, Furukawa K and Inoue H 1980 *The 8th International Liquid Crystal Conference* **E-11P** p 233
[14] Togashi S *et al* 1984 *SID Digest* (SID84) 324–325
[15] Lechner B J *et al* 1971 *Proc. IEEE* **59** 1566–1579
[16] Baraff D R *et al* 1980 *SID Digest* (SID80) 200–201
[17] Aota K, Kato Y, Hoshino K, Takahashi S, Irino M, Kikuchi M and Togashi S 1991 *SID Digest* (SID91) 219
[18] Takahashi K *et al* 1991 *International Display Research Conference* 247–250
[19] Grover C P and VanDriel H M 1980 *J. Opt. Soc. Am.* **70** 335–338
[20] Hashimoto N 1997 *OSA TOPS (Spatial Light Modulators)* **14** 227–249
[21] Goodmann J W and Silvestri M 1970 *IBM J. Res. Dev.* **14** 478
[22] Ichioka Y, Izumi M and Suzuki T 1971 *Appl. Opt.* **10** 403
[23] Moharam M G and Gaylord T K 1982 *J. Opt. Soc. Am.* **72** 1385–1392
[24] Botten L C, Craig M S, McPhedran R C, Adams J L and Andrewartha J R 1981 *Opt. Acta* **22** 1087–1102
[25] Nakata Y and Koshiba M 1990 *J. Opt. Soc. Am.* A **7** 1494–1502
[26] Swanson G J and Veldkamp W 1989 *Opt. Eng.* **28** 605–608
[27] Hutley M C 1982 *Diffraction Gratings* (New York: Academic)
[28] Gaskill D J 1978 *Linear Systems, Fourier Transforms, and Optics* (New York: Wiley)
[29] Shiono T, Kitagawa M, Setune K and Mitsuya T 1989 *Appl. Opt.* **28** 3434–3442
[30] Brinkly P F, Kowel S T and Chu C 1988 *Appl. Opt.* **27** 4578–4586
[31] Riza N A and DeJule M 1994 *Opt. Lett.* **19** 1013–1015
[32] Love G D, Major J V and Purvis P 1994 *Opt. Lett.* **19** 1170–1172
[33] Matic R 1994 Proc. *Proc. SPIE* **2120** 194–205
[34] Hashimoto N, Ogawa K, Morokawa S and Kodate K 1993 *IEEE Denshi Tokyo* **32** 67–70
[35] Ogawa K, Hashimoto N, Ito Y, Morokawa S and Kodate K 1993 *Proc. MOC/GRIN* 139–145
[36] Masuda S, Fujioka S, Honma M, Nose T and Sato S 1996 *Jpn. J. Appl. Phys.* **35** 4668–4672
[37] Milster T D, Wang M S, Li W and Walker E 1993 *Jpn. J. Appl. Phys.* **32** 5397–5401
[38] Hashimoto N and Milster T D 2001 *Proc. SPIE* **4342** 486–491
[39] Ohtaki S, Murao N, Ogasawara M and Iwasaki M 1999 *Jpn. J. Appl. Phys.* **38** 1744–1749
[40] Wang J Y and Markey J K 1978 *J. Opt. Soc. Am.* **68** 78–87
[41] Hashimoto N 1999 *Optical Apparatus* Japanese patent 2895150
[42] Tyson R K 1991 *Principles of Adaptive Optics* (Orlando: Academic)

[43] Iwasaki M, Ogasawara M and Ohtaki S 2001 *Tech. Digest* (Optical Data Storage Topical Meeting) 103–105
[44] Brat J 1997 *Appl. Opt.* **36** 8459
[45] Kodate K, Hashimoto A and Thapliya R 1999 *Appl. Opt.* **38** 3060–3067
[46] Baba N, Murakami N, Ishigaki T and Hashimoto N 2002 *Opt. Lett.* **27** 1373–1375
[47] Trimmer W 1997 *Micromechanics and MEMS* (New York: IEEE)

Chapter 3

On the use of liquid crystals for adaptive optics

Sergio R Restaino

3.1 Introduction

In the 17th century, Galileo Galilei pointed the first telescope to the heavens. In doing so, he did not just provide humanity with an enhancement of the eye but he also completely changed our position in the universe, psychologically and philosophically. Since then, astronomy has continued to change our view of the universe and our relationship to it. First, we went from being the centre of the universe to being in orbit around the centre of the universe. Then, instead of being the centre of the known universe, the sun was 'downgraded' to the status of an average star in the periphery of the galaxy. In the 20th century, it became clear that our galaxy is just one of the numberless galaxies in the sky. Since then the telescope has become not just a window on the universe, but a time machine, a high-energy physics laboratory and more. A large amount of our knowledge about exotic objects, and thus hard-to-reproduce physical situations, is gathered by astronomical observations. We are starting to get a better understanding of the physical phenomena involved relating to objects like neutron stars, black holes and quasars.

The single most important technological change that has allowed the dramatic increase in astronomical knowledge has been the ability to continue building more powerful telescopes. The ability of a telescope to resolve two close stars is directly proportional to the wavelength of the light used for the observations and inversely proportional to the diameter of the telescope itself (i.e. λ/D). This is why, in order to distinguish between ever closer pairs of stars (or to observe finer details on an astronomical object), we need to increase the diameter of the telescope for observations at a fixed wavelength of light. Of course, the other reason to make bigger telescopes is that they collect more light and this, in turn, allows us to observe dimmer objects. These are usually the most interesting, since very far away objects are dimmer than closer ones and these distant objects, that are far from us in time as well as in space, hold the physical information needed to address questions concerning the beginning and the end of the universe.

There are two major problems, however, to making increasingly larger optical telescopes. First of all, earth-based telescopes look at the skies through the earth's atmosphere. Isaac Newton, in his book *Opticks* published in 1730, investigated the random wandering of the images of stars seen through a telescope and correctly attributed this effect to the atmosphere: '...For the Air through which we look upon the Stars, is in perpetual Tremor...'. The earth's atmosphere is a turbulent medium that ultimately (and quickly!) limits the resolving power of the telescope to the intrinsic limit of the telescope's location. This limit is called *seeing* and is characterized by the parameter r_0 ('r nought'). This parameter is a statistical measurement of the strength of the turbulence at a given site. In this sense, it is similar to temperature, a single number which is a statistical property of the myriad atomic motions of an object. The seeing parameter, r_0, is measured in units of length and can be thought of as the size of a typical pocket of turbulence in the atmosphere. The second problem is related to the manufacturing process of large optical components that have micro-roughness of the order of a tenth, or better, of the wavelength of the light.

By its statistical nature, r_0 changes from site to site and, at the same site, it can change dramatically with time. At reasonable astronomical sites, average values of the seeing are around a few tens of centimetres. It is sobering to realize that the resolutions of large professional telescopes, several metres in diameter, are effectively the same as that of a well-equipped amateur observatory. Both amateur and professional astronomers have a common foe—the atmosphere. This is the reason why there are several projects for telescopes in space. However, in this case the cost of the project is such that only a very limited number of these can be carried out. This, of course, translates to only a few scientists being able to use such facilities.

Furthermore, a few years back, several scientists started to realize that the physical situation of a telescope trying to image a distant object through the turbulent atmosphere, is not that dissimilar to many other imaging and non-imaging problems. Let us look at some examples of other problems with similarity to what we have been discussing so far.

- Ophthalmologists are interested in taking high resolution imaging of the human cornea, *in vivo*, but the image quality is corrupted by the presence of the humor aqueous and the residual motion of the eye.
- In metallurgy, nowadays, the use of high-power lasers for cutting and welding is commonplace. However, during the procedure the laser interacting with the surface of the metal generates a cloud of plasma that behaves like the atmosphere, distorting the shape of the beam, resulting in less precise cuts or stitches.

- Propagation of laser beams through the atmosphere is also affected heavily by the turbulence. This results in, for example, reduced range for laser communication systems.

The list is, of course, not exhaustive but is enough to illustrate the need for ways to compensate for these defects that are generally referred to as 'aberrations'.

3.2   Adaptive optics: definition and history

One method of mitigating the effects of the atmosphere is to use a complex, closed-loop, system that dynamically compensates for the aberrations introduced. Such systems are known under the generic name of *adaptive optics* (AO). A schematic diagram of an AO system is shown in figure 3.1.

The three main components of such systems are a wavefront sensor, a corrective device and a control loop system. We will briefly examine these components later in the chapter. For the time being it is sufficient to point out that many solutions and researches have been devoted to each of these components and many different ways have been tested and selected in existing or under-development systems. However, given the character of

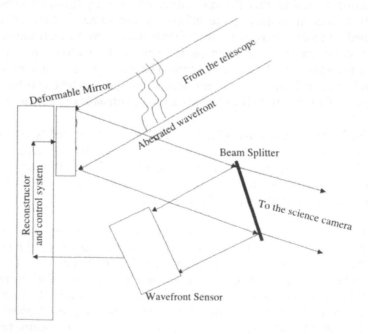

Figure 3.1. Block diagram of an AO system.

this book, we will focus our attention mostly on the use of liquid crystals as corrective elements for AO applications.

When and where was the first adaptive optics system invented and tested? According to ancient historians [1] during the Roman siege of Syracuse in 214 to 211 BC, Archimedes devised many war machines to stop and defeat the invading army of Marcellus. Among these war machines, legend lists the so-called burning mirrors. This was a system of mirrors that could be adjusted by Archimedes to focus the sun's rays and burn the ships of the Roman fleet. Accordingly, this was the first use of optics for military purposes.

In more recent times, the first idea of a system to mitigate the effects of the atmosphere was put forth by Babcock in 1953 [2]. While the Babcock system did not materialize, the idea was embraced by the military establishment in both the United States and the former Soviet Union. In the 1970s the first AO systems started to appear. In the US, most of the work was sponsored by the Advanced Research Program Agency (ARPA) and was declassified in 1991. Currently several AO systems are commissioned, under development or planned. Furthermore, in the past six or seven years, other applications for AO, as mentioned before, have started to emerge, ranging from medical imaging to laser fusion. In this climate, more and more research is being devoted to this complex problem and novel approaches are being tested. The driving reasons for these new approaches are many. One of the obvious reasons is cost and complexity. 'Classical' AO systems, and we will use this term for the systems that have been demonstrated so far, are expensive and very complex to operate. Furthermore, by the very nature of the systems, they are very dedicated instruments with little flexibility towards being able to be reconfigured, etc. It is in this framework that the use of liquid crystals for AO started to germinate several years ago. Liquid crystals offer the promise of many advantages. Let us briefly examine some of these advantages.

- Cost: Liquid crystal technology has advanced tremendously in the past decade or so, thanks to the investments in display technology. Many new compounds, addressing schemes, and clever cell design have risen from these investments and research. Because of the display industry efforts, it is also true that the cost for an LC device does not scale linearly with the number of channels, allowing for substantial savings when large numbers of corrective elements are considered in comparison with traditional technology.
- Complexity: The use of LCs, especially large format and high density of correction elements ones, may result in simplified AO systems from the control and use point of view.
- Lifetime and reliability: LC devices are very mature and reached a level where the lifetime is much larger than that of piezoelectrical devices, and the same goes for the reliability.

- No-moving parts: Extremely desirable especially for systems designed for space operations.
- Low power consumption: Traditional systems based on piezoelectric technology require large amounts of current in order to obtain significant strokes for correction. This is a problem for two reasons: (a) it drives the cost of the system up; (b) it makes the overall system physically larger and more complex. It finally makes it harder to use an AO system in environments like space.
- Large-format devices and large number of corrective elements: Taking into account the current display capabilities, in terms of size and quality, and recent advances in addressing schemes, it is realistic to say that large-format devices will be available soon. We will see later why dimension is also an important issue.
- Both transmissive and reflective devices are available, offering the opportunity of different optical designs for different applications. For example for high-accuracy systems, it may be useful to avoid the so-called pupil anamorphism. Such a term indicates that the projection of the circular pupil on a reflective device that is tilted by a certain angle is no longer a circle but an ellipse. This, in turn, creates an uneven mapping between points in the pupil and corrective elements on the device.

It would be disingenuous, however, to list all the possible advantages without mentioning the drawbacks and what can be done to overcome some of these. Of the possible drawbacks the most important ones are the following.

- Polarization issues: LC materials are usually birefringent materials, meaning that they are able to modulate only one polarization state. Two different approaches have been tested to resolve this issue. A description of these two techniques is given later in the chapter.
- Temporal bandwidth: This is an issue with conventional nematic materials. We will see later into the chapter how this problem can be addressed.
- Temperature sensitivity: Most LC materials will have a temporal and phase modulation response, that is a function of the temperature. This is probably the most delicate issue, and is heavily dependent on the specific LC material. We will address this issue towards the end of the chapter.

In order to understand how AO and liquid crystal devices operate, it is necessary briefly to introduce some basic image formation concepts and the effects of aberrations. This chapter is organized as follows: an introduction to image formation, followed by the effects of aberrations; an overall look at AO systems with a brief analysis of two wavefront sensors; and finally the use of liquid crystals for AO with an overview of what is the state-of-the-art in this arena.

3.3   Image formation: basic principles

Let us examine a single lens as a simple imaging system. The basic geometry of the problem is shown in figure 3.2.

The propagation of the light, under suitable and general conditions (see [3]), is governed by a linear differential equation of the form

$$\nabla^2 U(\vec{x}, t) - \frac{1}{c^2} \frac{\partial^2 U(\vec{x}, t)}{\partial t^2} = 0, \tag{3.1}$$

where we are assuming propagation in the vacuum away from charges and absorbers, and that one complex component of the electromagnetic field is representative of the phenomena under study, $U(\vec{x}, t)$. Justifications and discussion of the aforementioned assumptions can be found in reference [3]. The fact that the propagation is described by such an equation allows us immediately to state that each linear combination of solutions of equation (3.1) is still a solution. For most optical problems, the time dependence of the equation can be assumed as known in the form of $e^{(-i\omega t)}$. (This is equivalent to solving the PDE using a Fourier transform in the time domain.) In this way equation (3.1) becomes the so-called Helmholtz equation of the form

$$(\nabla^2 + K^2)\phi(\vec{x}) = 0, \tag{3.2}$$

where K is the modulus of the wavevector and its magnitude is $2\pi/\lambda$. This equation expresses the spatial evolution of the light. In optics, we are interested in what happens to a light wave when an obstacle, such as a screen with a hole or a lens, is placed in its path. In order to find the characteristics of the optical field at a distance from such obstacles, we can solve equation (3.2) using Green's theorem [4]. By choosing the right Green's function G, i.e. a spherical wave of the form $e^{(ikr)}/r$, we can write the Green's theorem in the following way:

$$U(\vec{X}) = \frac{1}{4\pi} \iint_\Sigma \frac{e^{ikr}}{r} \left[\frac{\partial U}{\partial n} - ikU \cos(\vec{n}, \vec{r}) \right] ds, \tag{3.3}$$

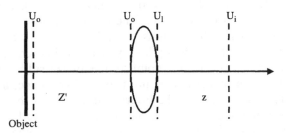

Figure 3.2. Geometry of the problem.

where Σ is the surface of integration and n is the normal to that surface taken positive in the outward direction. One of the basic assumptions that we will follow is that, unless otherwise noted, the source is a point-like source at infinity and we are dealing with semi-monochromatic light. The use of a point source at infinity is expressed mathematically by a complex function with a constant phase of the type $U(\vec{x}) = A(\vec{x})\,e^{i\phi(\vec{x})}$. The phase function $\phi(\vec{x})$ is the phase of the optical disturbance, $\phi(\vec{x}) = \vec{k} \cdot \vec{x}$, where $\vec{k}$ is the propagation vector of the wave, and $A(\vec{x})$ is the amplitude. The surfaces of equal phase are called the wave-front. The semi-monochromatic assumption is defined by stating that the bandwidth of the light used is smaller than the central wavelength in use, or $\Delta\lambda/\lambda \ll 1$. Furthermore, we assume that the observation plane is far enough so that the $1/r$ dependence can be dropped because it is negligible in comparison with the exponential term. A full treatment of diffraction theory is beyond the scope of this book; however, the interested reader can find exhaustive material in references [3] and [4], among others. After a long and tedious, but straightforward, manipulation of the basic Green's theorem we can find that under very general assumptions (such as, the source is a point-like source at infinity) the relationship between the known input optical disturbance, i.e. before an obstacle, and the output optical disturbance in a distant observation plane after the obstacle, is one governed by a Fourier transform:

$$U(\vec{X}) = \frac{1}{2\pi} \int_{-\infty}^{\infty} P(\vec{x})e^{[ik\vec{x}\vec{\xi}]}\,d\vec{x}, \qquad (3.4)$$

where $P(\vec{x})$ is a function that expresses the geometrical properties of the screen, lens, etc. This result is extremely powerful, and it is at the heart of the so-called Fourier optics discipline [4]. From equation (3.4) we can see that if our optical system is illuminated by a point source at infinity and in the absence of aberrations, the optical field distribution in the observation plane is the Fourier transform of the geometrical function describing the obstacle. Such function is called the pupil function. For example, in the case of a square hole in a screen, or a square lens, of length a, the pupil function is simply a function that is 1 between $-a/2$ and $a/2$ and 0 everywhere else. The Fourier transform of such function is the well known sinc function. This result can also be interpreted as a consequence of the Heisenberg uncertainty principle. The sinc function represents, in this case, the probability function for the photons to be located in a certain region of space after diffraction from the optical system. In this sense, this function indicates how the photons from a point source are spread by the diffraction process and form an image that is not a point-like image, as geometrical optics predicts. For this reason the Fourier transform of the pupil function illuminated by point source takes the name of point spread function (PSF). This is equivalent to the impulse response that is familiar in the linear system theory. To continue our similarity with linear system theory, we

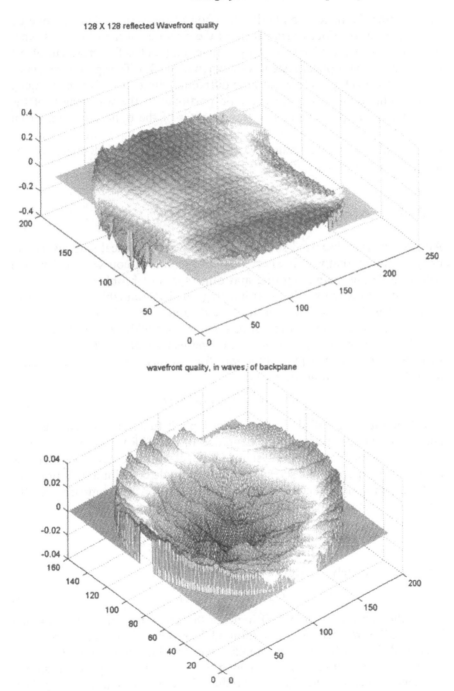

Figure 3.3. Measurement of the reflected wavefront quality of the device shown in figure 3.4. Above, back-plane only; below, device filled.

can examine the bandpass filtering function of the system. This is done by taking the reverse Fourier transform of the modulus of the PSF. This function takes the name of optical transfer function (OTF). The cross-cut of the OTF of a circular optical system is shown in figure 3.3. The spatial frequency at which the OTF goes to zero is the diffraction limit cut-off of the imaging system. This frequency is inversely proportional to the wavelength of the light in use and proportional to the diameter of the optical system. From this it is obvious that in order to obtain a higher spatial frequency cut-off, i.e. finer details on the image, we can either increase the diameter of the imaging system or use a shorter wavelength.

3.4   The effect of aberrations

All that we have discussed so far applies to 'perfect' optical systems, i.e. an imaging system that is only affected by diffraction. Of course, a real system will have design, manufacturing and other types of problems that will all contribute in degrading the final imaging quality attainable by the system itself. We will examine now how these degradations, aberrations, can be studied analytically and what can be done to cope with such problems.

Aberrations are departures from the perfect wavefront. It is convenient to include them in the pupil function, for reasons that will become clear immediately, in the following way:

$$\mathcal{P}(x, y) = P(x, y)\, e^{ikW(x,y)}. \tag{3.5}$$

The function $\mathcal{P}$ is called the generalized pupil function. We can use the generalized pupil function in the definition of the OTF to take into account the aberrations affecting the system. (Note that we are assuming that the aberrator is a pure phase screen and no amplitude contribution is present.) With this in mind, we can define an aberrated OTF and compare with the aberration free system. By using Schwarz's inequality

$$\left| \int XY\, d\xi \right|^2 \leq \int |X|^2\, d\xi \int |Y|^2\, d\xi, \tag{3.6}$$

we can immediately see that

$$|\text{OTF}_{ab}|^2 \leq |\text{OTF}_{dl}|^2, \tag{3.7}$$

where the subscripts ab and dl indicate the aberrated and the diffraction limited OTF respectively. Equation (3.7) states an important fact: the aberrated OTF will always be smaller than the diffraction limited one, i.e. the effect of aberrations on the imaging system will be to depress the spatial frequency content of the object and reduce the contrast. In practical applications it is convenient to define a parameter that is a measurement of how much an imaging system is aberrated. The parameter is the so-called Strehl ratio that we can define as the ratio of the volumes of the aberrated OTF over that of the diffraction limited OTF. This ratio is, of course, at maximum unity. From the relationship between OTF and PSF, and remembering Fourier transform properties, it is easy to see that another, totally equivalent, definition of Strehl ratio is the ratio of the peak intensity of the aberrated PSF over the diffraction limited one.

It is of practical use to expand the unknown aberration function W in a polynomial series. Several bases can be used; however, due to the recurrent circular geometry, it is more convenient to expand W in terms of a complete set of polynomials that are orthogonal over the interior of the unit circle. Many sets can be constructed to satisfy the aforementioned requirements. The set proposed by Zernike, that carries his name, has several appealing characteristics and has made this type of basis one of the most commonly used in optics. We define the Zernike polynomials as

$$Z_n^l(r\sin\theta, r\cos\theta) = R_n^l(r)\,e^{il\theta}, \tag{3.8}$$

where the polar coordinates have the customary relationship with Cartesian coordinates: $\rho = \cos(x)$ and $\theta = \tan(y/x)$. The aberration function can be expressed in terms of Zernike polynomials using the following expression:

$$W(r,\theta) = \sum_{n=0}^{K} \sum_{l=-n}^{n} C_{nl} R_n^{|l|}\,e^{il\theta}. \tag{3.9}$$

There are several advantages in using such polynomials, such as the fact that the first few Zernike polynomials resemble the standard basic aberrations, i.e. defocus, astigmatism, etc. Furthermore, it is easy to prove that the Strehl ratio can be expressed as the sum of the coefficients of the Zernike expansion of the wavefront.

So far we have seen how to describe the effects of aberrations on an imaging system. Let us turn our attention to how aberrations can be generated. In table 3.1, there is a list of the most common aberrations for a telescope. The aberrations are related to a temporal bandwidth, which represents the bandwidth of a corrective system that can compensate for such aberrations. It is important to note that these bandwidths are nominal and strongly dependent on many specific factors (environment, specific design etc.).

From this table, it is evident that the most important source of aberrations is the earth's atmosphere. The study of atmospheric turbulence is quite complex, especially because it is highly statistical in nature. Several authors have dedicated extensive work to the subject and we refer to some of them, [5] and [6] for example. We will limit ourselves to defining some basic parameters related to atmospheric turbulence. The most useful parameter is the so-called coherence diameter or Fried's parameter r_0. This parameter has the dimension of a length and can be thought of as the smallest area of atmosphere that generates only a tilt in the incoming wavefront and no

Table 3.1.

Source	Bandpass
Optical project	1–30 Hz
Manufacture	1–30 Hz
Thermal distortions	
mirrors	10^{-4} Hz
structure	10^{-3} Hz
Mechanical distortions	
mirrors	10^{-6} Hz
structure	varies
Wind	0.1–2 Hz (peak $= 1$ Hz)
Tracking errors	5–100 Hz
Atmospheric turbulence	10^{-2}–10^{3} Hz

higher-order aberrations. In a typical astronomical site and at visible wavelengths, i.e. 0.5 μm, r_0 is of the order of a few tens of centimetres!

3.5 Active and adaptive optics

There is not a consensus in the community about the use of the terms 'adaptive optics' and 'active optics'. For practical reasons, however, it is convenient to distinguish between the two on the basis of bandwidth and the use of wavefront sensor. We will define active optics as a system with a bandwidth that does not exceed 10 Hz and the wavefront sensor is not required. Very often, an active optics system is located directly in the entrance pupil of the system and is used to remove thermal and gravitational effects on the primary mirror of the telescope. Quite often, such effects can be modelled, and the deformation can be predetermined. In this case, the system is driven by a look-up table. On the other side, in order to remove aberrations that are random in nature and with bandwidths from several tens of Hz to 1 kHz, it is necessary to work in a relayed pupil and to use a wavefront sensor.

As we mentioned before, several AO systems are in use right now and several others are under development. All of these systems are based on a deformable mirror (DM), also referred to as a rubber mirror. The principle of using a mirror derives from the fact that the optical path difference (OPD), the quantity that we can really act upon and not the phase that is $2\pi/\lambda$ times the OPD, is expressed by

$$OPD = n\Delta z, \tag{3.10}$$

where n is the refractive index and Δz is the physical distance travelled by the wave. A DM can modulate the Δz.

Many different types of DM have been developed and tested. The basic components are a reflective surface that can be modified dynamically and some actuator system that will perform the deformation. Following is a list of some of the most common deformable mirrors that have been built and tested with some comments for each of the devices.

- Segmented mirrors: These mirrors are made by individual tiles that are attached to either single actuators (piston only) or three-element actuators (tip/tilt and piston). Used in the first solar adaptive optics system.
- Continuous surface mirrors: Usually a thin reflective surface that is mounted on discrete actuators. Several different types of continuous mirror have been developed using different kind of actuators (position, force and bending moment) and monolithic mirrors. The most commonly used DM.
- Membrane mirrors: Two main types of membrane mirrors have been developed and tested. In the first case a thin reflective membrane is positioned between a transparent electrode and a series of individual electrodes. More recently, advances in micro-machining technology have allowed the positioning of a thin membrane directly on electrodes. Usually these kinds of mirrors are more delicate than other DMs, and there are also physical constraints on the size attainable with this technology.
- Bimorph mirrors: This type of mirror consists of a glass or metallic mirror faceplate bonded to a sheet of piezoelectric ceramic. The ceramic is polarized normal to its surface. The bonding material between the mirror and the ceramic contains a conducting electrode. The exposed piezoelectric surface is covered with a number of electrodes.

One of the most important parameters to characterize a deformable mirror is its influence function. The influence function represents the physical fact that the action of a single actuator is not spatially limited to the extent of the actuator but affects a region around it. It is a function of mirror faceplate parameters such as thickness, modulus of elasticity, Poisson ratio, and of the location and distribution of the applied force. For many metallic materials, the influence function is found to be well approximated by a Gaussian or a super-Gaussian function (within 5% of accuracy). However, even if calculated, the influence function must be measured in order to take into account inhomogeneities, etc. In order to deform the surface of the mirror, it is necessary to use some form of actuators. Several different type of actuator have been developed. However, we can group the actuators in two broad classes: force actuators and displacement actuators. The most common materials for actuators are piezoelectric ceramics (PZT) and low-voltage lead–manganese–niobate (PMN). Some actuators have been manufactured using polyvinylidene fluoride (PVDF). By layering PVDF or stacking PZT, large strokes can be achieved with submicron accuracy. The high voltages needed to drive the stacks of PZT ($\sim$1 kV) often present serious design constraints. The development of magnetostrictive actuators reduces the requirements for high voltage (only a few volts are needed). However, the energy required to move the surface is the same, thus high currents (up to 1 A each) are required. Voice coil (solenoid) actuators are successfully used for tip/tilt mirrors mostly due to their large force and the displacement produced.

For low-bandwidth applications (active optics) a variety of different actuators can be used ranging from hydraulic to direct-acting d.c. motors.

Even though DM technology has been demonstrated and is currently widely used, there are many reasons to search for alternative solutions. In what follows, we will look briefly at two different ways of achieving phase modulation for adaptive optics.

3.6.1 Liquid crystal correctors

The idea of using liquid crystals (LC) as corrective elements dates back to the early 1980s. However, at that time the LC technology was not developed enough to produce usable devices. A dramatic change occurred in the early 1990s, especially due to the research and investments in this area related to display technology. At present, LC devices are available for use in laboratory set-ups and first telescope demonstrations.

The LC material (I will describe later the different type of materials available and of interest for adaptive optics) is sandwiched between two glass plates. The separation of the glass plates is maintained by spacers. On the glass plates is deposited a thin film of material that is a transparent electrode, usually indium tin oxide (ITO). The last layer is the alignment layer which is used to anchor the molecules. In conventional display technology, the two faceplates, with ITO and alignment films, are mounted perpendicularly to each other. The net result is that the spatial arrangement of the molecules is a spiral going from one extreme, the first face plate, to the orthogonal one on the other side. Because of this spiral arrangement, these devices are called twisted nematic. For phase modulation we need untwisted arrangements, where the face plates are parallel. There are also some additional reasons why the normal display technology of liquid crystals is not adequate for adaptive optics applications. Mainly, the optical quality of the face plates is not very high (see [7]) and additionally the single elements, pixels, are not controllable individually. Finally, the spacers are not located at the edge of the devices but are usually small spheres randomly spread throughout the surface of the device. This last issue is not a problem for display but may generate diffraction in a high-quality adaptive optical component. Referring to figure 3.3, the way that a liquid crystal device can modulate phase is related to the fact that the applied voltage will rotate the molecules, which are rod-shaped dipoles, to align with the field created within the cell. If we define $n_{\parallel}$ and $n_{\perp}$ as the ordinary and extraordinary components of the refractive index, then the phase modulation induced in one of the polarization components of the light going through the cell will be

$$\Delta\phi = \frac{2\pi}{\lambda} \int_{-d/2}^{d/2} [n(z) - n_\perp] \, \mathrm{d}z + \langle \Delta\phi \rangle_{\text{thermal}}, \qquad (3.11)$$

where the integral is taken over the thickness d of the cell, usually a few μm, and $n(z)$ varies from $n_\parallel$ to $n_\perp$. The thermal fluctuation term in equation (3.11), is usually negligible, of the order of 1.7×10^{-7} radians, for commonly used nematic materials. From equation (3.11), we can see that if we need more phase delay, we can increase the thickness of the cell; however, this will increase the response time of the cell. Another option would be to increase the optical anisotropy $\Delta n = n_\parallel - n_\perp$. Of course, for most AO systems, it is of interest to produce devices that can modulate the phase of unpolarized light. This can be achieved in two ways. Both techniques have been experimentally tested and devices have been built. The first approach is to build two devices and carefully align the two so that the optical axes of the two cells are orthogonal to each other. The other technique, first described by Love [8], consists of putting in optical contact a quarter-wave plate between the LC cell and a mirror. In this scheme, when light passes through the LC cell, one polarization state is retarded. The light then encounters the quarter-wave plate and reflects off the mirror and through the quarter-plate. This rotates the polarization of the light by 90°. The light then makes a second pass through the LC element, but this time the orthogonal polarization component of the light will be retarded.

At the beginning of the chapter we mentioned, among the advantages of using LC devices for adaptive optics, the issue of large formats. There are two main reasons why a large format is of interest. In astronomical applications one has to re-image the entrance pupil of the telescope, usually a few metres in diameter, onto the corrective device, usually few centimetres in dimension. Such a high level of demagnification causes a loss of usable field of view. The other reason lies in the use of LC devices with high-power lasers. It is desirable, in this type of application, to spread the overall power of the impinging laser beam over a larger surface in order to minimize damage to the device itself. A test device has been fabricated by Boulder Non-linear Systems in order to study the feasibility of large-format devices. The device in question is approximately $14 \, \text{cm}^2$ in clear aperture and has 128×128 corrective elements. The device is shown in figure 3.4. The basic parameters of the device are shown in table 3.2.

Of importance for adaptive optical applications is also understanding the overall optical quality of the device. This becomes a more important issue as the size of the device increases. As in all optical manufacturing processes the quality degrades with size. This degradation is usually some power law of the size, but since this power law is not an analytical law

Figure 3.4. New large-format LC device for wavefront control.

very often it needs to be established experimentally. In the case of the device in question we measured the optical quality of the substrate only and of the overall device, i.e. the device with all the layers and filled with LC material. Details of the measurements can be found in reference [9]; here we can report that the peak-to-valley (PV) wavefront error of the substrate only was measured to be $\lambda/13$ at 633 nm, and the PV value of the overall device was measured to be $\lambda/2$ at the same wavelength. This illustrates quite well the problems associated with the realization of high-quality, large format devices. Figure 3.3 shows a sample of the measured reflected wavefronts through the device with the substrate only (above) and filled device (below).

Another area of interest for LC correctors is the geometry of the corrective elements and also the possibility of producing influence functions that are not just 'top-hat' functions, i.e. corrective elements that are not only piston term modulators, but can also apply tip and tilt. Extensive studies in

Table 3.2.

Cell thickness	10 µm
Material	Rodic RDK-01160
Birefringence	$\Delta n = 0.20$ at 25 °C and $\lambda = 589$ nm
Clearing point	94 °C
Melting point	−25 °C

this area have been carried out by several groups, especially in Russia, the Netherlands and the UK [10–12]. First of all they were able to test a system using a differential resistivity on each corrective element. This generated a gradient in voltage that in turn generates the wanted tip and tilt modes [11]. The other interesting area of research is the ability to control the device in a modal fashion more like a membrane mirror [12]. Other areas involve the construction of variable lenses using LC devices. This is accomplished by using circular segments as corrective elements, and by applying a voltage profile to the device that is compatible with the phase front that one wants to generate. In this way the LC device can be compared with a Fresnel lens. While, from the correction point of view, there are not specific advantages of one configuration versus the other, it is quite important when one is interested in completing systems by matching the appropriate wavefront sensor to the appropriate device. In section 3.8 we will give two examples, one of a zonal sensor (Shack–Hartmann) and one of a modal sensor (phase diversity).

3.6.2   What kinds of LC are of interest?

There are several thousands of compounds which are classified as liquid crystals. A classification of all these compounds is usually done on the basis of the physical mechanism that induces the mesophase. The first class is called lyotropic, in which the influence of solvents is the physical mechanism. The lyotropic family is the largest. It is very common in nature (soap, cell membranes, etc.) but is of no interest for adaptive optics. The second class is called thermotropic, where thermal processes are responsible for the mesophase. The thermotropic class is composed of three large families of compounds, nematic, cholesteric and smectic. This last family contains several different classes like smectic A and C (usually indicated as SmA and SmC), and recently a few more classes have been identified— SmB, E, G, H, J and K. The two types of liquid crystals of interest to us are the nematic and a couple of smectic classes like SmA and C. (Usually the smectic compounds are also called collectively ferroelectric.) A simple diagram that compares the nematic and ferroelectric materials with equivalent optical components, i.e. waveplates, is shown in figure 3.5. The diagram also compares the average switching times of the two kinds of LC materials.

From this diagram, we can see how the nematic LC can be easily used to replace a conventional DM. The main drawback in using nematic materials is the speed at which they can be switched. Research is still on-going in this area. I will also illustrate some of the options that may be available in the future for the use of nematic LC material. The use of ferroelectric LC materials presents a more challenging approach since it is not straightforward to use them to replace a DM. However, new materials with different characteristics are available now that may allow us to use ferroelectric

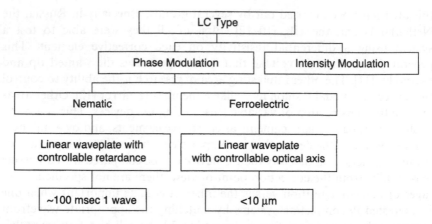

Figure 3.5. Schematic comparison between LC materials and equivalent optical elements.

materials in a fashion similar to nematic materials. Of course, the obvious advantage of using ferroelectric materials resides in their intrinsically faster response time. Currently, our main experimental effort is concentrated in the nematic arena since it is the most mature technology. Among others, we are using a 127-element device built by Meadowlark Optics in Boulder, Colorado. The device is composed of two orthogonal layers of nematic material that permits its use for unpolarized light. The aberrated PSF obtained during lab tests, and the same PSF after correction using the 127 elements, are shown in figure 3.6.

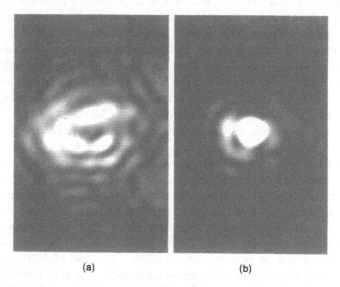

Figure 3.6. (a) Aberrated PSF and (b) corrected PSF using the 127-element Meadowlark device.

From this figure, it is evident that a high level of correction is possible. The Strehl ratio before and after the correction was of 27% and 75%, respectively. However, the future for nematic LC materials resides in the use of dual-frequency material. The dielectric permittivities of all liquid crystals vary with the frequency of the applied field when $\nu = 10^8$ Hz. In the low-frequency range, hundreds of Hz to tens of kHz, $\varepsilon_\parallel$ and $\varepsilon_\perp$ are usually constant. However, for certain materials, at low frequencies, $\varepsilon_\parallel$ changes which leads to a change in sign of $\Delta\varepsilon$. ($\Delta\varepsilon = \varepsilon_\parallel - \varepsilon_\perp$.) For most of these materials $\Delta\varepsilon$ is positive at the low end of the frequency range of the applied voltage, and negative at the high end. The frequency at which $\Delta\varepsilon$ reverses sign is called the crossover frequency ν_c. The sign reversal of $\Delta\varepsilon$ can be used to orient the liquid crystal either with optical axis parallel ($\Delta\varepsilon > 0$) or normal ($\Delta\varepsilon < 0$) to the direction of the electric field by selecting the frequency of the applied voltage. For ferroelectric materials the most important novelty has to do with new materials that exhibit high-tilt angle and more or less continuously switchable angle. Up to now, conventional SmC materials have been considered inappropriate for analogue spatial light modulators (SLMs) due to their intrinsically binary nature, i.e. for a device using a material with a given switching angle, the device could only be switched between the initial state (no field applied) and the final tilt angle.

3.7   Characterization and control of nematic LC devices

There are three main factors that control the rapidity of switching of the molecules of a liquid crystal device: viscosity and elastic constant, thickness of the layer, and control voltage. We will examine, briefly, each of these factors.

3.7.1   Viscosity and elastic constant

Choosing a liquid crystal with low viscosity to increase the temporal rate of the cell response usually leads to an increase of the control voltage. Since most liquid crystal materials of interest, as a rule, possess a small optical activity, a high degree of reorientation of the molecules is required for the same variation of the phase delay. This effect can be characterized approximately by the reaction factor $\gamma_1/\Delta n$ [13], where $\Delta n = n_\parallel - n_\perp$ is, as usual, the optical anisotropy and γ_1 is the viscosity coefficient. Viscosity can be decreased by increasing the temperature according to $\gamma_1 \simeq \exp(A/kT_0)$ where A is an activation energy [14], and k is the Boltzmann constant. However, in this case the elastic constant K_{11} is decreasing also. The parameter of interest is thus γ_1/K_{11}. It can be established experimentally that the ratio γ_1/K_{11} decreases monotonically with an increase in temperature to some optimal temperature T_{opt}. Above T_{opt} the elastic constant drops more significantly than the effective viscosity and this results in a slower decay time [15]. Furthermore, experience shows that the spatial gradient in temperature needs to be maintained within less than 0.1 °C in order to avoid spatial degradation of the wavefront transmitted through the layer [16].

3.7.2 Thickness of the layer

The speed of the molecules' switching rate is determined by the thickness of the layer as $1/d^2$. Therefore, thin cells are preferable for fast phase modulation. However, this approach presents two problems. The first is a small depth of phase modulation as can be seen through equation (3.11). The second is a higher probability of break-down or short-circuit. In the real substrates the liquid crystal molecules' surface alignment is seldom uniaxial due to the surface roughness, defects or irregular alignment procedures. The 'virtually inactive' thickness is defined by the quality and preparation of the substrates. Under conditions of strong anchoring and good surface quality, the thickness of the partially disordered surface layer is reduced. On the other hand, a large virtually inactive thickness is expected for weak anchoring and rough surface alignment. Usually the virtually inactive thickness is estimated at ~0.3 μm for each surface, and an estimate of ~0.5 μm for the thickness of the inactive volume per cell is thus obtained.

3.7.3 Control voltage

Reorientation can be induced by both a static or an alternating electrical field. In the case of static fields the current must flow through the electrodes, so electrodes' processing becomes important. The electric field in the liquid crystal layer becomes inhomogeneous and is determined by anisotropy conditions. Double layers are formed at the electrodes, to cope with this problem, which in turn decreases the field in the cell, so a higher applied voltage is needed to reorient the layer as in the case of alternating fields. It has been found that the development of the double layers takes several seconds, so they can be completely neglected for alternating fields.

Nematic liquid crystals have been used as phase retarders for wavefront shaping for a number of years, for example, see references [9–12, 17, 18]. In order to control and characterize a LC device, we must understand its physical behaviour. Let us start with equation (3.11). In this equation $n(z)$ is related to the rotation of the liquid crystal molecules by

$$n(z) = \frac{n_{\parallel}n_{\perp}}{[n_{\perp}^2 \cos^2(\theta(z)) + n_{\parallel}^2 \sin^2(\theta(z))]^{1/2}}, \quad (3.12)$$

where $\theta(z)$ is the distribution of molecular rotation along the z axis that is also the axis of propagation of the light. Such distribution can be expressed as a superposition of spatial modes:

$$\theta(z, t) = \sum_j \theta_j(t) \cos\left[(2j - 1)\frac{\pi z}{d}\right].$$ (3.13)

For small rotation angles, of less than $50°$, the first-order mode dominates, i.e. $\theta(z, t) = \theta_l(t) \cos(\pi z/d)$. Then equation (3.12) can be expanded in a power series, and, taking into account only the small rotations, we obtain

$$\langle n(z) \rangle = n_\| \left(1 - \frac{(n_\|^2 - n_\perp^2)/n_\perp^2}{4} \theta_l^2 + \cdots\right),$$ (3.14)

where $\langle n(z) \rangle$ is the index of refraction integrated over the length of the cell. The temporal response of the nematic liquid crystal birefringence is highly complex and not completely understood; for example see reference [19]. However, it is possible to achieve some level of understanding in order to design some control algorithm. The dynamic behaviour of the rotation state of the liquid crystal axis is described by the Ericksen–Leslie equation [19, 20]:

$$\frac{\partial}{\partial z}\left[(K_{11}\cos^2\theta + K_{33}\sin^2\theta)\frac{\partial\theta}{\partial z}\right] - (K_{33} - K_{11})\sin\theta\cos\theta\left(\frac{\partial\theta}{\partial z}\right)^2$$

$$+ (\alpha_2\sin^2\theta - \alpha_3\cos^2\theta)\frac{\partial v}{\partial z} + \frac{\Delta\varepsilon E^2}{4\pi}\sin\theta\cos\theta$$

$$= \gamma_1\frac{\partial\theta}{\partial t} + I\frac{\partial^2\theta}{\partial t^2}.$$ (3.15)

The terms K_{11} and K_{33} are the splay and bend Frank elastic constants [20], and γ_1 is the Leslie rotational viscosity coefficient. I and v are the inertial moment of the molecules and the flow velocity, respectively, and both quantities are usually negligible compared with the elastic and viscosity constants. Finally, E is the applied electric field. There is no known solution of equation (3.15). However, with suitable simplifications approximate solutions can be found that are useful, such as

$$\theta_l^2(t) = \frac{1}{2}\left(\frac{V^2 - V_0^2}{\frac{2}{3}V^2 + KV_0^2}\right)\left(1 + \tanh\frac{t - t_0}{\tau_1}\right).$$ (3.16)

The term V_0 is the so-called threshold voltage,

$$\tau_1 = \frac{\gamma_1 d^2/K_{11}\pi^2}{(V^2/V_0^2) - 1},$$ (3.17)

$$K = (K_{33} - K_{11})/K_{11}.$$ (3.18)

From these results we can now write the phase retardance in terms of the applied voltage. Assuming that the time t is large enough that the phase retardance will settle to a static value, i.e. we are not looking around the transient time, and assuming $V \gg V_0$, we obtain

$$\Delta\phi = \Delta\phi_{max}\left[1 - \frac{1}{4}\left(\frac{n_{\parallel}^2}{n_{\perp}^2} + \frac{n_{\parallel}}{n_{\perp}}\right)\left(\frac{V^2 - V_0^2}{\frac{2}{3}V^2 + KV_0^2}\right)\right]. \qquad (3.19)$$

With these results, and other approximate results [20, 21], it is now possible to devise a controlling algorithm for the phase retardance of a liquid crystal device to be used as an adaptive optics corrective element. Experimental results on LC devices can be found in the literature; for example see references [17] and [21].

The control of a liquid crystal device can be achieved using four different techniques: amplitude control, transient method, pulse method and dual frequency control.

3.7.4   Amplitude control

This is the simplest method and employs the amplitude variation of the rms (root mean square) value of the applied voltage. The dependence of $\Delta\Phi(U)$ is monotone with the voltage U. For example, for a cell thickness of $5\,\mu m$, an applied voltage ranging from 1 to $5\,V$, and with $\gamma_1 \approx 10\,cP$ and $K_{11} = 10^{-6}$ dyn, one obtains a value for the 'turn on' time of approximately $25\,\mu s$ and for the 'turn off' time of $0.25\,s$. Of course these values are material specifics and many other factors, like alignment layer, manufacturing defects, etc., contribute to the measured values.

3.7.5   Transient method

The idea of the transient nematic effect (TNE) [22] is to use the fast decay time due to the small relaxation angle which in turn is due to highly deformed liquid crystal directors. To better understand the principle let us describe the operational steps involved in using such method. A relatively high a.c. voltage is applied to the liquid crystal cell. As a result, almost all the molecules are aligned by the electric field approximately orthogonal to the substrate surfaces, except in the boundary layers. When these highly deformed directors start to relax, i.e. the voltage is removed completely, the directors undergo free relaxation. In order to stop the directors' motion a voltage is applied to the cell.

3.7.6   Pulse method

The use of a bipolar rectangular control voltage allow us to drive a modulator by varying the period-to-pulse duration ratio $q = T/\tau$, where T is the period duration and τ is the pulse length. The amplitude of the mth harmonic is given by [23]

$$a_m = \frac{2V_0}{\pi m} \sin\left(m\pi\frac{\tau}{T}\right). \tag{3.20}$$

When $T < 0.01\,\text{s}$ the birefringence depends on the acting voltage value and does not depend on the sign of the applied voltage at that instant. Thus, variation of the parameter q leads to a change in the harmonics' amplitude and, consequently, to a variation of the phase modulation depth of the liquid crystal layer.

3.7.7 Dual frequency

We have already described very briefly the concept of a dual-frequency control in the previous section. The two-frequency addressing scheme was first proposed for dynamic-scattering-type liquid crystal displays. The operation principle consisted of the transition from a conductivity anisotropy regime to a dielectric anisotropy induced alignment regime at the cut-off frequency [24]. For adaptive optics purposes, instead, the use of dual frequency control is to increase the rapidity of a liquid crystal wavefront corrector.

3.8 Wavefront sensing techniques

In order to close the loop of an adaptive optical system the other essential element is the wavefront sensor. Many different techniques and ideas have been developed in the past few years. For the sake of brevity and simplicity, we will examine briefly only two wavefront sensing techniques.

The basic idea for wavefront sensing comes from the techniques used in optical testing, i.e. the testing of optical components. However, there are several meaningful differences between the two applications (table 3.3).

There are two general philosophies for detecting and expressing the wavefront.

- The wavefront is expressed in terms of optical path difference (OPD) over a small area. This approach is called zonal.

Table 3.3.

AO requirements	Optical testing requirements
High temporal frequency	Low temporal frequency
High spatial resolution	Usually low spatial resolution
Unknown aberrations	Range of aberrations usually known

- The wavefront is expressed in terms of coefficients of a global polynomial expansion over the entire pupil of the system. This approach is called modal.

While the choice between modal and zonal depends on the application, both techniques have been used for atmospheric correction. Usually, however, preference is given to the modal approach if low-order aberrations are of interest, and vice versa the zonal approach is used for high-order aberrations correction. Some hybrid approaches have also been proposed. It can be shown that with a zonal approach, with correction zones of characteristic length r_s, the Strehl improvement can be written as

$$S_Z = \exp[-(2\pi)^2]0.008\left(\frac{r_s}{r_0}\right)^{5/3}. \qquad (3.21)$$

Similarly, one can write the Strehl improvement based on the number M of modes corrected, as

$$S_M = \exp[-0.2944M^{0.866}]\left(\frac{D}{r_0}\right)^{5/3}. \qquad (3.22)$$

For identical AO correction, i.e. identical Strehl corrections, the number of modes

$$N_M = 0.92\left(\frac{D}{r_s}\right)^{1.93} \qquad (3.23)$$

and the number of zones

$$N_Z = 0.78\left(\frac{D}{r_s}\right)^2 \qquad (3.24)$$

are roughly equivalent. Both expressions represent the number of degrees of freedom of the system.

3.8.1   Shack–Hartmann wavefront sensor

The Shack–Hartmann wavefront sensor [25] derives from the classical Hartmann test [26] developed for testing large optical surfaces. The Hartmann test belongs to a category of optical tests called screen tests. The basic idea is to place an opaque mask, with holes arranged in a certain pattern, behind the optical element being tested. The result is an array of spots. With proper calibration, the position of each spot is a measurement of the local wavefront tilt at each hole, and thus a description of the overall optical quality of the system under study. Shack placed lenses in the holes, which increased the light-gathering capabilities of the system and increased the sensitivity of the sensor. In practice, one measures Δx and Δy on an array sensor, placed on the focal plane of lenslet array. These measurements are directly proportional to the slope of the wavefront. There are several parameters to keep in mind when designing a Shack–Hartmann wavefront sensor. Let us start by pointing out that the calibration is very important, i.e. in order to measure the spot displacement one has to have a reference for an unperturbed system. Second, the noise characteristics of the sensor are directly related to the detection process of the single sub-aperture (i.e. the single lens in the lenslet array) of the sensor. To illustrate this situation let us look at a practical example. Let us suppose that we have an entrance pupil of 3 m, with an r_0 of 10 cm. The photon flux from a star of apparent visual magnitude m is

$$F = 0.7 N \tau D^2 \Delta \lambda 10^{-0.4m} \tag{3.25}$$

where $N = 10^4$ photons/(s cm^2 nm) for an A0 star at 550 nm, τ is the system transmittance (we assume 70% transmittance), D is the pupil diameter and $\Delta \lambda$ is the bandpass in nm. With $m = 9$, we obtain a flux of 1.1×10^8 photons per second. We will use half of our signal for the wavefront sensing and we will use an exposure time for the detector of 5 ms (in order to 'freeze' the atmosphere). The Hartmann sensor will have 30×30 subapertures and will be perfectly matched to the detector (i.e. central peak of the diffraction pattern equal to one pixel on the detector), and we will need *only* 2×2 pixels for the detection. With these numbers, we obtain an average of 77 photons per pixel. The expression of the signal-to-noise ratio (SNR) in terms of the photon flux is given, under certain conditions, by

$$\text{SNR} = \frac{kFQ}{\sqrt{kFQ + \langle n \rangle}} \tag{3.26}$$

where k is a factor included to take into account losses not included in the transmission term, Q is the quantum efficiency of the detector (for CCD we will use 75%) and $\langle n \rangle$ is the sum of all the contributors to noise (i.e. thermal effects, electronic read-out noise, etc.). Using a benign estimate for k of 99% and assuming a perfect detector (i.e. $\langle n \rangle = 0$), we obtain an SNR of 7 that is good enough for detection. Of course, in a realistic system the situation is always worse. This creates the need for bright reference stars that can be used for wavefront sensing, where bright means usually a visual magnitude smaller than 7. The distribution of 5th and 6th magnitude stars in the sky is far from uniform and creates serious problems for wavefront sensing. Two solutions have been proposed and tested. The first is to use sub-apertures that are larger than r_0. By using this method the light efficiency can be greatly augmented at the expense of other parameters, each sub-aperture no longer subtending only a tilt, and the image is thus not a diffraction pattern but a speckle pattern. The measurements of Δx and Δy have a higher noise content, thus the wavefront estimate is less accurate. The other approach is the use of laser-guided stars. The concept is to use a powerful laser to generate a reference beam in the atmosphere. Two approaches are possible: Rayleigh scattering and sodium layer excitation. The first approach has been demonstrated at the Starfire Optical Range (SOR) at Kirtland Air Force Base in New Mexico, USA. A laser is propagated from the telescope up to 20 km in the atmosphere, where Rayleigh scatter will generate a reference beacon. One of the advantages of this technique is that it moves with the pointing of the telescope. One of the disadvantages, however, is that it does not sample the entire column of atmosphere and the correction is only partial. In the sodium layer approach a more powerful laser is propagated up to the sodium layer, around 90 km from the ground, and excites the sodium molecules that will re-emit light at the familiar 517 and 518 nm wavelengths.

3.8.2 Phase diversity (curvature sensing)

A completely different approach to wavefront sensing is related to the general approach of measuring wavefront aberrations by analysing the focal pattern of an optical system [27–29]. To illustrate how such a system works let us examine a simple optical system, a single lens, with only one aberration, astigmatism. One way of detecting the aberration is to look at the image formed by the lens of simple object, a cross hair. The image will have the arms of the cross hair focused in two different planes, so by scanning the focal volume of the optical system we can detect aberrations. Let us generalize this concept for any aberration and find an analytical expression that we can use for the data reduction. We assume that we have a monochromatic point source at infinity that is described by the expression

$$U(\vec{r}) = I(\vec{r})^{1/2}\, e^{i\phi(\vec{r})} \tag{3.27}$$

with the customary notation that the intensity I is the square modulus of the amplitude. This wave is, of course, a solution of the Helmholtz equation [equation (3.2)]. Now we make the so-called slowly varying envelope approximation, i.e. we assume that U varies slower in the propagation direction z than in the perpendicular planes x and y. This implies, mathematically, that we neglect the second derivative of U with respect to z and retain only the first-order derivative in z. In other words, we state that we approximate the spatial evolution along the z axis by $U(\vec{x}, z) = U(\vec{x})\, e^{ikz}$. When we use this approximation in the Helmholtz equation we obtain

$$\left(-i\frac{\partial}{\partial z} + \frac{\nabla^2}{2k} + k\right)U = 0. \tag{3.28}$$

This is the so-called parabolic wave equation. It is worth noting that equation (3.28) is mathematically identical to the Schrödinger equation. This is a result of the fact that the Schrödinger equation is also the result of a slowly varying envelope approximation for a wave packet. We can manipulate equation (3.28) by multiplying the left-hand side by U^* and the complex conjugate of equation (3.28) by U, and subtracting the two expressions. The result, after rearranging terms, is

$$-k\frac{\partial I}{\partial z} = \nabla \cdot (I\nabla\phi) = \nabla I \nabla \phi + I\nabla^2 \phi. \tag{3.29}$$

This is the so-called transport equation for intensity and phase. First of all, let us note that equation (3.29) is nothing else than an energy conservation law (compare with $-\partial \rho/\partial t = \nabla \cdot (\rho \vec{v})$ where ρ is the mass or charge density and $\vec{v}$ is the flow velocity). If we measure the intensity pattern at different focal planes, i.e. we measure $\Delta I/\Delta z$ as an approximation for $\partial I/\partial z$, we can solve equation (3.29) and obtain an estimate of the wavefront. Note that since equation (3.28) is a transport equation, the measurements can be made in both regions around the pupil plane of the focal plane, when the system is illuminated by a point source. If the system is illuminated by an extended source it is preferable to work around the pupil plane, since the phase retrieval in the image plane will be complicated by the image itself in this case. If we are around the pupil plane an approximation often used, even though it is not rigorously true, is that the intensity in these planes is uniform, in this case the gradient of the intensity will be zero, or negligible compared with the phase excursions, and equation (3.29) can be re-written as

$$-k\frac{\partial \log I}{\partial z} = \nabla^2 \phi. \tag{3.30}$$

There are two advantages in using equation (3.30): first, it is a well known Poisson equation, and the second advantage derives from the deformation law of a bimorph mirror that follows a Poisson equation for the voltage applied. In this way an almost one-to-one connection can be made between the wavefront sensor and the corrective element.

The analysis of the wavefront-sensing techniques carried out so far is limited strictly to point sources. Completely different approaches need to be taken when dealing with extended sources. In the case of a Shack–Hartmann sensor, because each lenslet will form an image of the object, the customary technique of finding the centre of gravity of the focal spot in order to measure the displacement in x and y cannot be used. Nevertheless, it is possible to use autocorrelation techniques in order to find the local displacements. However, this is quite costly in terms of temporal bandwidth of the system. In the case of phase diversity, measurements in the image plane are still possible but many frames have to be used to reconstruct an average object phase that can be used to invert the problem into the pupil plane and thus retrieve the aberrator phase profile.

3.8.3 Putting it all together

We have briefly looked at the main ingredients for an AO system. Now we can put things together and give an example of an LC-based AO system,

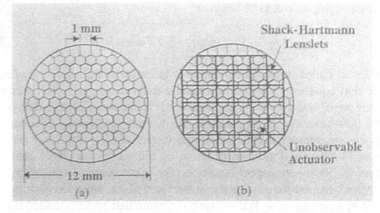

Figure 3.7. Meadowlark Optics multisegmented device: (a) arrangement of segments and (b) arrangement of Shack–Hartmann lenslets.

with some experimental results. (More details of this system can be found in references [17] and [21].) The LC device being used has 127 hexagonal corrective elements, and was built by Meadowlark Optics. The preliminary step is to identify the wavefront sensing scheme to be used. In this experimental set-up the choice was a 6×6 Shack–Hartmann wavefront sensor. There are several parameters to keep in mind when using a Shack–Hartmann sensor in connection with a liquid crystal device. For example, since the aforementioned LC is a piston-only device, it is not possible to match the lenslet array with the pixels of the corrector on a one-to-one basis. The scheme chosen is to match each lenslet to a group of pixels as shown in figure 3.7.

Figure 3.7 also shows that roughly four elements of the Meadowlark device are under each subaperture. In this way we can circumvent the problem that Shack–Hartmann sensor cannot sense the piston mode. Even with this arrangement some unobservable elements, as shown in figure 3.7, may exist and possibly cause instability. In order to maintain stability in the feedback control algorithm, we must limit the degrees of freedom in the LC device surface to force the unobservable actuators to move in concert with their neighbours.

The first step, in actuating the loop, is to reconstruct an estimate of the wavefront error by using the measured Shack–Hartmann data. Next a polynomial fit is performed using the first 28 Zernike modes. The reasons for using a polynomial fit are many-fold. One significant reason is that a polynomial fit acts as a high spatial frequency filter, thus smoothing the data from high frequency components. The use of only a limited number of modes in the expansion is one way of limiting the degree of freedom of the system. The use of 28 modes was reached through trial and error. Because the chosen wavefront sensor produces phase gradient measurements on a square grid, the control algorithm must provide for interpolation of the reconstructed phase onto the hexagonal grid of the LC device. The Zernike modal expansion is used to accomplish this in the following way:

$$\Phi_{\text{hex}} = [Z_{\text{hex}} \quad Z_{\text{sq}}^{-1} \quad H] \begin{bmatrix} \nabla\phi_{x11} \\ \nabla\phi_{x12} \\ \vdots \\ \nabla\phi_{y11} \\ \nabla\phi_{y12} \\ \vdots \end{bmatrix}. \tag{3.31}$$

In equation (3.31), H is the standard least-squares wavefront matrix for a square-array Shack–Hartmann sensor, Z_{sq} is a Zernike expansion matrix sampled on a square grid, and Z_{hex} is a Zernike expansion matrix sampled on a hexagonal grid. The product of these three matrices forms an interpolated control matrix, which we can implement in high-speed real time hardware. The vector $\nabla\phi$ represents the wavefront gradient errors as measured by the Shack–Hartmann sensor. The subscripts x and y indicate the x and y direction gradients respectively. The numbers following the x and y subscripts are indexes that refer to the position within the two-dimensional grid of Shack–Hartmann lenslets for that particular gradient. At this point Φ_{hex} represents an array of the changes in phase retardance that we would like to apply to the elements of the device to compensate for the wavefront errors measured by the wavefront sensor. Such a scheme has been tested in the laboratory environment and on a telescope. An example of experimental results obtained with a dual-frequency device is shown in figure 3.8. The star used for this test is β Delphini, a binary star with a difference in visual magnitude between the companions of $\Delta m = 3$. The closed loop bandwidth, i.e. the 3 dB rejection, of the system was of 40 Hz.

Figure 3.8. Experimental results using a dual-frequency liquid crystal device. The star is β Delphini (a binary star). The left image shows the open loop frame and the right image shows the closed loop frame.

3.9   Conclusions

In this chapter, we have reviewed the basic principles of image formation and how aberrations affect the final image. We have also discussed the most common strategy to deal with the effects of dynamically changing aberrations: adaptive optics, and how these techniques have developed in the past three decades. Finally, we have reviewed the reasons for the use of liquid crystals as correctors. Currently there are several international conferences that are held on a regular basis dealing with the fast-changing state-of-the-art technology in the area of adaptive optics. As we have mentioned throughout the chapter, it is becoming increasingly evident that these type of solution may have a wider range of applicability than just mitigation of atmospheric turbulence. The other aspect that needs to be stressed is that the core technologies that have a direct bearing on this technique, i.e. computer systems and LC device technology, are fast changing. The number of researchers, and thus the number of different approaches and ideas, is also increasing. All this conspires to a very dynamic and quick changing arena. On the other hand, the classical approach is very costly and complex and few systems are available and producing routine scientific data. This is probably one of the most compelling reasons to research for alternative techniques that will allow a further spreading of these technology and thus warrant a larger and more routine use of adaptive optics.

References

[1] Simms D L 1977 Archimedes and the burning mirrors *Technol. Culture* **18**(1) 1–24
[2] Babcock H W 1953 The possibility of compensating astronomical seeing *Publication ASP* **65** 229–236
[3] Born M and Wolf E 1980 *Principles of Optics* (Cambridge: Cambridge University Press) 6th edn
[4] Goodman J W 1968 *Introduction to Fourier Optics* (New York: McGraw-Hill)
[5] Tatarski V I 1961 *Wave Propagation in a Turbulent Medium* (New York: McGraw-Hill)
[6] Roddier F 1981 The effects of atmospheric turbulence in optical astronomy *Progress in Optics* **19** 333
[7] Love G D, Fender J S and Restaino S R 1995 Adaptive wavefront shaping with liquid crystals *Optics and Photonics News*, 16–20 October
[8] Love G D 1993 Liquid crystal modulator for unpolarized light *Appl. Opt.* **32**(13)
[9] Restaino S R, Martinez T, Andrews J R and Teare S W 2002 On the characterization of large format LC devices for adaptive and active optics *Proc. SPIE* **4889**
[10] Love G D *et al* 1994 Liquid crystal prisms for tip-tilt adaptive optics *Opt. Lett.* **19**(15)
[11] Naumov A, Loktev M, Gralnik I and Vdovin G 1998 Cylindrical and spherical adaptive liquid crystal lenses *Proc. of Laser and Optics conference, Saint Petersburg*
[12] Loktev M, Vdovin G, Naumov A, Saunter C, Kotova S and Guralnik I 2002 Control of a modal liquid crystal wavefront corrector *Proc. 3rd International Workshop on*

the use of Adaptive Optics for Medicine and Industry ed S R Restaino and S W Teare (Albuquerque)

[13] Clark M G 1998 Dual-frequency addressing of liquid crystal devices *Microelectronic Reliability* **21**(6)

[14] Blinov L M 1983 *Electro-Optical and Magneto-Optical Properties of Liquid Crystals* (New York: Wiley)

[15] Wu S T 1986 Phase retardation dependent optical response time of parallel-aligned liquid crystals *J. Appl. Phys.* **60**(5)

[16] Restaino S R 1999 Liquid crystal technology for adaptive optics *Proc. SPIE* **3635**

[17] Restaino S R *et al* 2000 On the use of dual frequency nematic material for adaptive optics systems: first results of a closed-loop experiment *Optics Express* **6**(1) 2–6

[18] Love G D 1997 Wave-front correction and production of Zernike modes with a liquid crystal spatial light modulator *Appl. Opt.* **36** 1517–1524

[19] Chandrasekhar S 1992 *Liquid Crystals* (Cambridge: Cambridge University Press)

[20] Dorezyuk V A, Naumov A F and Shmal'gauzen V I 1989 Control of liquid crystal correctors in adaptive optical systems *Sov. Phys. Tech. Phys.* **34**

[21] Dayton D, Browne S, Gonglewski J and Restaino S R 2001 Characterization and control of a multielement dual-frequency liquid crystal device for high-speed adaptive optical wave-front correction *Appl. Opt.* **40**(15) 2345–2355

[22] Wu S T and Wu C-S 1988 Small angle relaxation of highly deformed nematic liquid crystals *Appl. Phys. Lett.* **53**(19)

[23] Vasil'ev A A, Naumov A F, Svistun S A and Chigrinov V G 1988 Pulse control of a phase corrector liquid crystal cell *J. Tech. Phys. Lett.* **14**(5)

[24] Chang T S and Loebner E E 1974 Crossover frequencies and turn-off time reduction scheme for twisted nematic liquid crystal displays *Appl. Phys. Lett.* **25**(1)

[25] Shack R B and Platt B C 1971 *J. Opt. Soc. Am.* **61**

[26] Malacara D 1977 *Optical Shop Testing* (New York: Wiley)

[27] Teague M R 1983 Deterministic phase retrieval: a Green's function solution *J. Opt. Soc. Am.* **73**

[28] Restaino S R 1992 Wavefront sensing and image deconvolution of solar data *Appl. Opt.* **31**(35)

[29] Roddier F, Roddier C, Graves J E and Northcott M J 1997 *Astrophys. J.* **443**

Chapter 4

Polymer-dispersed liquid crystals

F Bloisi and L Vicari

4.1 Introduction

Both polymers and liquid crystals (LCs) belong to a class of 'condensed matter' materials, sometimes called 'complex fluids' or 'soft matter' [1–3], whose physical and chemical properties and behaviour have been extensively studied only in relatively recent years (20th century). Within soft matter, characterized by large molecules, constraints typical of solid state coexist with thermal fluctuations dominating the fluid state. Unlike gases or liquids, soft materials do have some shape (polymers) or internal organization (LCs), but unlike 'hard materials' they strongly respond to small external mechanical (polymers) or electric (LCs) disturbances.

In the simplest form, polymers consist of long 'chains' or 'necklaces' of identical units, the 'monomers', flexibly bonded to each other. More complex polymeric structures can use different kinds of monomers (copolymers) or contain branched chains or crosslinks. The polymerization reaction, required to produce the polymer starting from the monomer, often requires a sort of 'activation' or 'initiation', which can be obtained adding a chemical 'initiator' or by means of exposition to high-energy photons (UV light). Some polymeric materials, such as rubber, were known and used since the 19th century, but their properties and characteristics have been extensively studied only starting in the 20th century. The peculiar structure of polymers allows production of materials with very interesting mechanical properties such as high resistance coupled with high flexibility.

LCs [4–7] are highly anisotropic (calamitic, discotic or lath-like) weakly coupled molecules showing one or more liquid 'mesophases' (i.e. intermediate phases) [8] in which their properties (partial or no positional order, partial orientational order) are intermediate between those of an isotropic liquid (neither positional nor orientational order) and those of a crystal (both positional and orientational order). The result is a material that couples most of the mechanical properties of a liquid (high fluidity, inability to support shear, etc.) with some electromagnetic properties (high

148

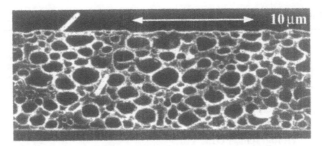

Figure 4.1. Morphology of a PDLC film prepared with emulsion technique. From: H Ono and N KaWatsuki 1994 *Jpn. J. Appl. Phys.* **33** L1778.

electrical, magnetic and optical anisotropies) of a crystal. LCs were known since the end of the 19th century but first applications in electro-optic devices appear in the middle of the 20th century. In such applications the most attractive properties concern the possibility of controlling the LC optical anisotropy by means of an electric or magnetic field, while the liquid behaviour is an undesired characteristic.

To find more useful materials some authors have used LCs in compound materials such as droplets of nematic material floating in an isotropic liquid [9] or nematic LC confined in micrometre-sized cavities within a solid [10, 11].

Fergason in 1984 [12, 13] and Doane and co-workers a few years later [14, 15] introduced a new class of composite materials constituted by small (order of magnitude 0.1–100 μm) droplets of LC embedded in a polymeric film. The currently most used term is PDLC (polymer dispersed liquid crystal) even if sometimes the term NCAP [16] (nematic curvilinear aligned phase or encapsulated liquid crystal) is used, mainly for 'emulsion type' PDLCs [17–22]. Generic terms such as LCPC (liquid crystal–polymer composites) or PNLC (polymer–nematic liquid crystal) are seldom used.

The typical aspect of a PDLC observed using a scanning electron microscope is shown in figure 4.1 [23].

PDLC, and more generally dispersions of LC molecules in a homogeneous material, have been a subject of interest in the study of surface effects because they may be predominant with respect to bulk effects, due to the confinement of LC molecules into small volumes [24–28]. Moreover, PDLCs are of interest for their nonlinear behaviour, since their optical properties can be controlled by the optical electric field [29–31].

Often different terms are used to specify that a chiral [PDCLC (polymer dispersed chiral liquid crystal) or PDCN (polymer dispersed chiral nematic)] [32, 33] or a ferroelectric [PDFLC (polymer dispersed ferroelectric liquid crystal)] [34, 35] LC is used. Sometimes a different term is used to specify a different preparation technique or operation mode of the PDLC film: [HRPLC (homeotropic reverse-mode polymer LC)] [36] or HPDLC (holographic PDLC) [37–40]. Finally, for the sake of completeness, we must

Table 4.1. List of abbreviations.

ADA	Anomalous Diffraction Approach
BSO	$Bi_{12}SiO_{20}$ photoconductive crystal
EA-SLM	Electrically Addressable SLM
GDLC	Gel–glass Dispersed Liquid Crystals
HRPLC	Homeotropic Reverse-mode Polymer-LC
HPDLC	Holographic PDLC
H-PDLC	Holographic PDLC
ITO	Indium Tin Oxide (In_2O_3:Sn)
IR	Infra Red
LC	Liquid Crystal
LCD	Liquid Crystal Display
LCPC	LC–Polymer Composites
MIM	Metal Insulator Metal
NCAP	Nematic Curvilinear Aligned Phase (or encapsulated liquid crystal)
N–I	Nematic-to-Isotropic (phase transition)
NLC	Nematic Liquid Crystal
OA-SLM	Optically Addressable SLM
PDCN	Polymer Dispersed Chiral Nematic
PDLC	Polymer Dispersed Liquid Crystal
PDLC-LV	PDLC Light Valve
PDLC/LM	PDLC Light (intensity) Modulator
PDLC/PM	PDLC light Phase Modulator
PDCLC	Polymer Dispersed Chiral Liquid Crystal
PDFLC	Polymer Dispersed Ferroelectric Liquid Crystal
PIBMA	Poly-(IsoButyl-MethAcrylate)
PIPS	Polymerization Induced Phase Separation
PMMA	Poly-(Methyl-MethAcrylate)
PNLC	Polymer-NLC
PoLiCryst	Polymer-Liquid Crystal
P-PIPS	Photoinitiated PIPS
PSCT	Polymer Stabilized Cholesteric Texture
PSLC	Polymer Stabilized Liquid Crystal
PVA	Poly-Vinyl-Alcohol
PVF	Poly-Vinyl-Formal
RGA	Rayleigh–Gans Approximation
SIPS	Solvent-Induced Phase Separation
SLM	Spatial Light Modulator
TFT	Thin Film Transistor
TFEL	Thin Film Electro Luminescent
TIPS	Temperature-Induced Phase Separation
T-PIPS	Thermally-initiated PIPS
UV	Ultra Violet
n-CB	alkylcyanobiphenyls (e.g. 5-CB, 6-CB, etc.)

mention some compound materials having more or less PDLC-like electro-optic properties: polymer stabilized cholesteric texture (PSCT) containing cholesteric LC and a very low amount of polymer [41, 42]; polymer–liquid crystal (PoLiCryst) obtained sandwiching an LC between two rough polymer surfaces [43]; transparent solid matrix imbibed with a liquid suspension of anisotropic particles [10]; optically nonabsorbing porous matrix filled with an LC [11, 44]; cholesteric LC–polymer dispersion [45]; cholesteric LC–polymer gel dispersion [46]; etc.

The introduction of PDLCs reaches the aim to couple the peculiar mechanical properties of a polymeric film (flexibility and high mechanical resistance) to the peculiar electro-optical properties of LCs (electrically controllable high optical anisotropy), allowing the realization of many new applications (e.g. flexible displays, privacy windows, projection displays, sensors, etc.).

Within this chapter, section 4.2 is devoted to explaining the working principle and the preparation techniques of a PDLC film.

Section 4.3 is a survey of the main physical problems involved in the study of PDLCs: namely the effects of combined electric and elastic forces on LC molecules in confined volumes and the light scattering by small aniso-tropic particles.

Section 4.4 shows the numerical or approximate approaches usually followed in the description of electro-optical properties of PDLCs.

Finally section 4.5 is devoted to the description of several PDLCs applications.

Table 4.1 will be useful for readers unfamiliar with some usual abbreviations.

4.2 PDLC preparation techniques

The term 'PDLC film' usually means a solid but flexible film (the polymer) containing a more or less large number of cavities (the 'droplets') filled with an LC. This definition includes a lot of different materials having different properties, sometimes arranged for specific applications. In this section, devoted to discussing general preparation techniques, we do not refer to any specific application; however, to fix our ideas, we briefly describe the working principle of a 'classical' PDLC film application: the 'PDLC light shutter' (figure 4.2).

Let us assume that (i) the polymer is isotropic and nonabsorbing; (ii) the droplets are almost spherical in shape, and are randomly and uniformly distributed within the polymer; (iii) the droplet size is comparable with the wavelength of visible light; (iv) the LC is in the nematic mesophase, is non-absorbing and its ordinary refractive index, n_o, equals the refractive index of the isotropic polymeric medium, n_p; (v) the film is sandwiched between

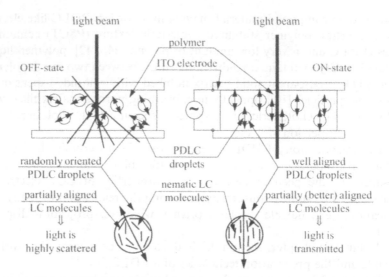

Figure 4.2. Schematic representation of a PDLC light shutter in the OFF-state (left) and when a low frequency electric field is applied across it (ON-state, right). Double arrows are a schematic representation of the droplet director (see section 4.3.1).

two transparent electrodes allowing application of a (static or low frequency) electric field orthogonally to the film surface; (vi) the dielectric anisotropy of the LC is positive so that the torque generated by an electric field tries to align the LC molecules to the direction of the field; (vii) no ordering has been induced, during PDLC film preparation, in droplet or LC molecule orientation.

If no electric field is applied (OFF-state) the nematic LC is partially ordered, but the order does not extend outside droplets: the LC molecules inside each droplet are partially aligned to each other while droplets within the PDLC film are randomly oriented (figure 4.2, left). The optical properties of each droplet are those of a uniaxial material (the optical axis is represented by double-headed arrows (↔) in figure 4.2), but the optical properties of the whole PDLC film are those of an isotropic inhomogeneous material. Due to the difference between the refractive index of the polymeric film and the refractive indices of the LC inside droplets (refractive index mismatch), light passing through a PDLC light shutter in the OFF-state is highly scattered so that the film appears opalescent (milky).

The amount of scattered light is related to a large number of parameters concerning the light beam (wavelength, incidence angle, polarization state) the operating conditions (temperature, value and waveform of the applied electric or magnetic field) the component characteristics (polymer and LC refractive indices and dielectric constants, LC elastic constants) the PDLC configuration (droplet shape, size, uniformity and distribution). A detailed

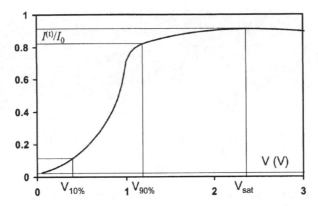

Figure 4.3. Typical behaviour of the light intensity transmitted by a PDLC film.

discussion will be developed later in this chapter, but it can be easily understood that light scattering certainly vanishes if there is no refractive index mismatch. The application of a suitable electric field across the film (ON-state) aligns all LC molecules, thus reducing to zero, at least for light with normal incidence, the refractive index mismatch. As a consequence the PDLC light shutter in the ON-state (figure 4.2, right) behaves as an anisotropic material, perfectly transparent for a normally impinging light beam.

When the electric field is removed the LC molecules recover their initial orientations and the PDLC film returns to its highly scattering behaviour. Figure 4.3 shows transmitted light intensity versus the applied electric field for a PDLC film.

It must be noted that the transition from the OFF- to the ON-state is continuous so that we can use a PDLC film as a PDLC light intensity modulator (PDLC/LM, figure 4.4).

Some parameters, useful to characterize a PDLC device, have been defined [47, 48] as follows.

- *Contrast ratio* is generally defined as the ratio $I_{max}^{(t)}/I_{min}^{(t)}$ between maximum and minimum transmitted light at normal incidence. To increase the ON-state transmittance the polymer refractive index, n_p, must be matched to the droplet ordinary refractive index, n_{do}. To reduce the OFF-state transmittance the LC optical anisotropy, i.e. the difference between LC extraordinary and ordinary refractive indices $\Delta n = n_e - n_o$, must be as large as possible. Note that PDLC light transmittance is angle dependent and, for off-axis incidence, it is polarization dependent. Moreover light scattering, and therefore contrast ratio, is wavelength dependent. Contrast ratio is a very important parameter in all applications, but it must be used with care since often, in normal operation conditions, light does not impinge orthogonally to PDLC film surface.

(a)   (b)   (c)

Figure 4.4. A PDLC light modulator prepared in our laboratory. The PDLC film is $20\,\mu m$ thick and the low frequency (50 Hz) electric field is $E = 32\,V/20\,\mu m$ in the ON-state (c). From: L Vicari and F Bloisi 2003 *Optics and Lasers in Engineering* **39** 389.

- *Driving voltage* is generally defined as the lowest voltage, V_{sat}, at which the transmitted light reaches its maximum value I_{max}. Sometimes the two values $V_{90\%}$ and $V_{10\%}$ are used instead of V_{sat}, V_f being the voltage at which transmitted light is $fI_{max}^{(t)}$. A low-frequency a.c. voltage (frequency order of magnitude 10^1–10^3 Hz) is usually used to avoid conductivity effects.

- *Response (rise and decay) times.* Rise time is usually defined as the time needed for transmitted intensity to reach 90% of the saturation value, when the driving voltage is applied. Similarly decay time is defined as the time needed for transmitted intensity to fall to 10% of the saturation value, when the driving voltage is removed. Even if submillisecond response times have been reported [49] response time orders of magnitude are typically 1–10 ms for rise and 10–100 ms for decay times. Dual-frequency LCs [50, 51] have been used to speed up decay time. A dual-frequency LC is a mixture behaving as a positive LC for low-frequency applied fields (below kHz order of magnitude), and behaves as a negative LC for high-frequency applied fields (above kHz order of magnitude). A low-frequency electric field is used to switch ON the PDLC film, while a high frequency electric field is used to rotate LC molecules away from aligned direction. It has also been found [32] that a small amount of cholesteric dopant sensibly reduces the decay time.

- *Hysteresis* is usually defined as the difference between the voltages required to reach half light transmission intensity ($I_{50\%}^{(t)}$) during switching ON ($V_{50\%}^+$) and switching OFF ($V_{50\%}^-$): $\Delta V_{50\%} = V_{50\%}^+ - V_{50\%}^-$. It has been generally observed that alignment is better achieved for decreasing than for increasing values of the electric [52, 53] or magnetic fields [54]. It has been suggested that hysteresis is due to the difference in response time of bulk and surface LC layer [55].

- *Haze* is defined as the fraction of light scattered out of a cone of 5° of full aperture, with respect to the total transmitted light. This is an important parameter since in a PDLC film non-transmitted light is scattered and not absorbed. When observing an object through an ON-state PDLC film some scattered light seems to come from dark areas thus producing a hazy image. Moreover it must be taken into account that an ON-state PDLC film has an anisotropic behaviour. Since such an off-axis haze effect is polarization dependent, it can be reduced (at the cost of reducing the overall transmission ratio) by introducing a polarizing sheet [56].
- *Charge holding ratio* is generally defined as the ratio of the voltage across a pixel at the end and at the beginning of a refresh cycle. Each pixel of a display, in usual operation, is only addressed for a short time (about 60 μs) each refresh cycle (about 100 ms). Therefore the applied voltage must not sensibly decrease when the electric circuit is open. The effect of voltage decay can be reduced, in active matrix configurations, by introducing a capacitor for each pixel.

Another characteristic to be taken into account in realizing PDLC-based devices is time stability. Stability of the mechanical and electro-optical properties of the PDLC film over time is important in practical applications. The main problems are connected to increasing of polymer crosslinks with time. Also photochemical stability must be taken into account, to avoid deterioration of polymer or LC.

4.2.1 General preparation techniques

Many recipes for the preparation of PDLC films can be found in the literature. However, all of them are based on very few preparation techniques. Moreover all preparation techniques can be grouped in two main classes depending on whether one starts with an emulsion of the LC in the polymer (or the corresponding monomer) or with a single-phase solution of LC and polymer (or monomer). In the first case, the 'emulsion techniques', the LC droplets are formed in the liquid phase while in the second case, the 'phase separation techniques', the droplets are formed later, during film solidification.

We must observe that other compound materials containing LCs have been developed and used, but strictly they cannot be considered as PDLCs even if they have similar electro-optical properties. Some of these materials can be considered as precursors and others as further development of PDLCs, so we will briefly consider them at the end of this section.

Several parameters, influencing the PDLC film behaviour, must be taken into account in the choice of preparation technique:

- *Droplet size.* Both its average value and its uniformity affect light scattering.

- *Droplet shape*. The shape of the droplet influences the LC orientation while its more or less uniform distribution within the PDLC film may influence light scattering.
- *Phase contamination*. Polymer or monomer molecules trapped within the LC (i.e. within droplets) influence the refractive indices of the droplets while LC molecules trapped in the polymer (i.e. outside droplets) influence the refractive index of the surrounding medium, and both reduce the refractive index mismatch in the OFF-state.
- *Film thickness*. Once the applied voltage has been fixed, the electric field across the PDLC film is proportional to its thickness, so in large area devices both its average value and its uniformity must be taken into account.
- *Film ageing*. Time variation of the chemical and physical properties of the PDLC film are important in applications.

4.2.1.1   Emulsion technique

The emulsion technique has been introduced by Fergason [12, 13] and is generally recognized as marking the birth of PDLCs, even if the term PDLC was introduced later.

An emulsion-type PDLC [sometimes, for historical reasons, referred as NCAP (encapsulated liquid crystal)] is obtained by preparing an emulsion of LCs in an aqueous solution of a water-soluble polymer, and letting the solvent evaporate. Polyvinylalcohol (PVA) is mainly used in the preparation of emulsion PDLCs; the emulsion is usually produced by rapidly stirring the components. Droplet size is determined by the characteristics of the emulsification processes (i.e. stirring speed and duration). Due to volume change during solvent evaporation, spherical droplets within the emulsion result in elliptical droplets, with major axes in the plane of the film, in the emulsion-type PDLC film.

Since polymerization is produced by solvent evaporation, the PDLC film must be prepared by placing the liquid emulsion over a single transparent conducting electrode [usually a glass plate or a plastic film with an ITO (indium tin oxide) coating] and letting it polymerize. The film can be covered with the second transparent conducting electrode only after polymerization is complete. Due to this procedure, there may be technical difficulties in obtaining good adhesion of the second conducting stratum on the already polymerized PDLC film. For the same reason there may be some difficulties in obtaining a PDLC film having the desired uniform thickness.

A method of obtaining good uniformity in film thickness for small samples is to place the required quantity of the emulsion at the centre of a spinning (over 1000 rpm) ITO-coated glass substrate.

LCs generally have very low solubility in water and in water-soluble polymers so that it is possible to assume that no phase contamination occurs during PDLC preparation. On the contrary, water-soluble polymers are hygroscopic and this can sensibly reduce the lifetime of an emulsion-type PDLC film.

The most used materials are PVA [17, 55, 57, 58], a mixture of PVA and glycerol [59] or polyurethane latex [58].

Emulsion-type PDLC recipe [60]: A solution of 15 wt% PVA in distilled water is prepared at room temperature (23 °C). The solution is heated at 100 °C to dissolve PVA completely and then is allowed to cool at room temperature. LC (ZLI-2061 or ZLI4151-100 or E7 from Merck), 60 wt% in the composite film, is dispersed in the PVA/water solution and the mixture is stirred at 5000 rpm for 5 min ON and OFF every 15 s with a small propeller blade. The mixture is creamy white since it contains air, so it is maintained at rest for 20–24 h to degas. Finally the emulsion is spin-coated onto an ITO-coated glass at 1500 rpm for 90 s to produce a uniform thickness sample.

4.2.1.2 Phase separation techniques

The phase separation technique was introduced by Doane and co-workers [14, 15] and marks the start of usage of the term PDLC. In the phase separation technique the PDLC is obtained starting with a homogeneous, liquid, single phase mixture containing both the LC and the polymer (or monomer, or pre-polymer). During polymer solidification almost all LC molecules are 'expelled' from the polymer (phase separation) and aggregate in droplets which remain embedded in the polymeric film.

The phase separation can be induced in several ways:

4.2.1.3 Polymerization-induced phase separation (PIPS)

The monomer or pre-polymer is mixed with the required amount of LC in a single-phase liquid. The liquid is placed between two transparent conducting electrodes and the polymerization process is started, by heating (thermally initiated PIPS, or T-PIPS) or illuminating (photo-initiated PIPS, or P-PIPS) the sample, depending on the polymer type.

T-PIPS type PDLC recipe:
Components (wt%):
E7 (45.0), Epon-828 (27.5), Capcure 3-800 (27.5) [61]
E7 (33.3), Epon-828 (33.3), Capcure 3-800 (33.3) [62]
E7 (41.0), Epon-828 (11.0), Capcure 3-800 (28.0), MK-107 (20.0) [63]

Preparation:
The components are put together and stirred to obtain a homogeneous mixture. The mixture is then centrifuged to remove air. The mixture is sandwiched between ITO-coated glass plates with Mylar spacers. The

sample is heated (typically at 50–100 °C) and held at this temperature (typically 1–24 h) to allow polymerization.

4.2.1.4 *Temperature-induced phase separation (TIPS)*

The LC is mixed with a melted thermoplastic polymer, the liquid is placed between two transparent conducting electrodes and phase separation is induced by polymer solidification, obtained by cooling the sample at a controlled rate.

4.2.1.5 *Solvent-induced phase separation (SIPS)*

A solution is prepared with the polymer, the solvent and the required amount of LC. The liquid is placed over a single transparent conducting electrode. After the solvent has evaporated, thus inducing droplet formation, a second transparent conducting electrode is placed over the PDLC film.

Both PIPS and TIPS techniques allow the PDLC film to be prepared directly between the two required transparent electrodes and this has two advantages. The first is that there is good adhesion of the PDLC film to both electrodes so that the final device is mechanically more robust than emulsion-type PDLCs. The second advantage is that it is possible to obtain a more uniform and controlled PDLC film thickness just by adding a small amount of 'spacers' (often glass spheres or cylinders) to the liquid single-phase mixture. For small PDLC samples, known thickness Mylar films are used as spacers.

Droplet size mainly depends on the LC volume fraction in the mixture and on the phase separation reaction rate [63]: a slow reaction produces larger droplets than a fast reaction, since LC molecules have more time to aggregate. Size uniformity depends on the LC-to-polymer mass density ratio [63]: a density ratio approaching unity prevents droplet sedimentation and avoids the formation of droplet clusters.

Phase contamination [64] may constitute a problem since it affects the refractive index of the droplets and of the surrounding medium and changes the clearing point: it has been shown [64, 65] that the clearing point of E7 (a eutectic mixture by Merck) is lowered by 3–7 °C with respect to the temperature of the nematic-to-isotropic (N–I) transition for pure LC. Moreover the LC molecules trapped in the polymer do not contribute to the droplet formation altering the LC/polymer volume fraction while the monomer trapped in the droplets may polymerize during usage, thus altering PDLC characteristics.

The most used materials in T-PIPS preparation are epoxy resins such as Epon-828 or Epon-165 [66–69] together with a hardener (e.g. the polymer-captane Capcure 3-800). Often an aliphatic compound (MK-107) [70] is used to obtain a better refractive index match. P-PIPS is commonly obtained starting with a UV-curable adhesive [68, 71, 72] (e.g. Norland NOA-65).

Finally TIPS [67, 73, 74] and SIPS [62, 75–77] techniques are usually based on vinyl polymers such as PMMA, PVF and PVA (for planar anchoring) or PIBMA (for homeotropic anchoring).

LCs used are often alkylcyanobiphenyls (e.g. 5-CB, 6-CB) or alkyloxycyanobiphenyls (e.g. OCB), or eutectic mixtures (e.g. E7, E8, E9 by Merck).

4.2.2 Special techniques and materials

We briefly describe here some PDLCs obtained with special preparation techniques or operated in non-standard modes. A detailed discussion of their behaviour requires a more complete understanding of PDLCs and is postponed in the section devoted to PDLC applications.

4.2.2.1 Reverse-mode PDLCs [36, 73, 78–81]

Reverse-mode, i.e. transparent in the ON-state and scattering in the OFF-state, PDLC have been obtained in several ways: for example, by using an LC with negative dielectric anisotropy in droplets with radial configuration [78], taking advantage of the dependence of LC behaviour on applied electric field frequency [79], or using an anisotropic polymeric matrix [81].

4.2.2.2 Haze-free PDLCs [45, 56, 81, 82]

A PDLC film in the ON-state is anisotropic so light transmitted has a high angular dependence. In large-area devices the result is an unwanted haze effect. Such effect can be reduced adding a polarizing sheet [56] to selectively absorb light contributing to an off-axis haze effect, or using an anisotropic polymer [81] to reduce angular dependence of transmitted light, or using a dichroic dye doped LC [82].

4.2.2.3 Temperature-operated PDLCs [65, 83]

In 'classical' PDLC operation the non-scattering state is obtained by applying an electric field, but a non-scattering state can also be obtained by heating the PDLC film so that LC inside droplets undergo N–I phase transition. If the refractive index of the polymer is equal to the refractive index of the isotropic LC the whole PDLC film behaves as a homogeneous isotropic material.

4.2.2.4 Phase modulation [70, 84]

The presence of LC droplets in the PDLC film affects not only the intensity of transmitted light, but also its phase. Light scattering is due to refractive index mismatch, but requires droplets with size comparable with visible light wavelength. Producing a PDLC with smaller droplets allows the use of the PDLC film as an electrically controllable phase modulator (PDLC/PM).

4.2.2.5   Holographic PDLCs [37–40, 85]

If light used to generate a P-PIPS PDLC is spatially periodic (obtained by interference of two coherent light beams) droplet density distribution follows the same spatial periodicity. This allows the recording of a holographic grating within the PDLC (HPDLC: holographic PDLC) which can be switched ON and OFF by means of an electric field.

4.2.2.6   Scattering PDLC polarizer [59, 86–89]

Usually LC droplets in the PDLC film are randomly oriented, but it is possible to obtain a more ordered distribution by stretching the film after polymerization [59, 86, 87]. This gives a markedly anisotropic behaviour to the PDLC film so that it acts as a polarization-selective scattering object. A modulable scattering light polarizer can be obtained by applying an electric field parallel to the PDLC film surface [88, 89].

4.2.2.7   Gel–glass dispersed liquid crystals (GDLC) [90, 91]

In recent years a technique using gel–glass to trap LCs in a homogeneous matrix has been proposed. GDLC is prepared with a phase separation technique, stirring together a liquid silicon alkoxide, a metal alkoxide and an LC and letting the sol–gel process proceed. During gelling the solubility of LCs in the matrix decreases and droplets of LCs are formed in the gel–glass matrix.

4.3   The physics involved in PDLCs

The internal structure of a PDLC film may be very complex, but for many applications it can be sketched as an optically non-absorbing inhomogeneous material composed of an isotropic solid phase (the polymer) containing almost spherical droplets filled with an anisotropic liquid (the LCs).

The main physical problems involved in this schematic view are the determination of the distribution of the LC molecules inside the droplets, both in the unperturbed state or when an external constraint (e.g. electric field) is applied, and the electric and optical behaviour of an inhomogeneous material (the whole PDLC film). Therefore, in studying PDLCs, it is of fundamental importance to understand the mechanical behaviour of

LCs in confined volumes and the light scattering from small anisotropic particles.

The aim of this section is to describe these physical problems in a general form, while approximations and numerical approaches will be discussed in section 4.4.

4.3.1   LC in confined volumes

In a simple and intuitive approach we can imagine elongated LC molecules, typically used in PDLCs, as 'rod-like'. It is therefore possible to associate a 'molecular director', i.e. a unit vector $\hat{l}$, to each molecule. Since LC molecules contained in a small volume are not randomly oriented, the average position of the molecular director defines a local privileged direction, the 'nematic director' $\hat{n} = \langle \hat{l} \rangle$. The aim of the elastic continuum theory applied to an LC is to determine the 'director field' $\hat{n}(r)$. This is usually accomplished by minimizing the total (electric, magnetic and elastic) free energy density. The electric and magnetic contribution have their usual forms, while the elastic contribution, the 'distortion free energy', follows from the generalized Landau–de Gennes theory which combines the Landau phase transition theory, extended by de Gennes to LC transitions, with orientational elasticity theory.

4.3.1.1   The order parameter

Except in the isotropic phase, LCs are characterized by local anisotropy: one or more physical properties (e.g. the polarizability) must be expressed in the form of a tensor. The eigenvectors of any tensorial quantity define a privileged local orthonormal reference frame $(\hat{n}, \hat{n}', \hat{n}'')$. The largest eigenvector or the nondegenerate one, since LCs often behave as locally uniaxial material, defines a privileged direction given by the nematic director $\hat{n}$.

Based on the ideas of Landau–de Gennes theory it is possible to describe an LC by means of a tensor-order parameter taking into account both the local reference frame and the degree of order. This is accomplished by looking at any tensorial quantity $\mathbf{T}$ (e.g. the magnetic susceptibility tensor or the dynamic dielectric tensor) and taking into account its anisotropic normalized (dimensionless) part,

$$Q_{ij} = G(T_{ij} - \tfrac{1}{3} \operatorname{Tr}(\mathbf{T})\delta_{ij}), \tag{4.1}$$

where δ_{ij} is the Kronecker symbol and $\operatorname{Tr}(\mathbf{T}) = T_{11} + T_{22} + T_{33}$ is the trace of tensor $\mathbf{T}$.

The 'tensor order parameter' $\mathbf{Q}$ is real, symmetric and of zero trace and can be written, in the privileged local reference frame, in the diagonal form

$$\mathbf{Q} = \begin{pmatrix} -(Q_1 + Q_2) & 0 & 0 \\ 0 & Q_1 & 0 \\ 0 & 0 & Q_2 \end{pmatrix} \tag{4.2}$$

$$= \begin{pmatrix} \frac{2}{3} S^{(u)} & 0 & 0 \\ 0 & -\frac{1}{3}(S^{(u)} - S^{(b)}) & 0 \\ 0 & 0 & -\frac{1}{3}(S^{(u)} + S^{(b)}) \end{pmatrix} \tag{4.3}$$

with

$$S^{(u)} = -\tfrac{3}{2}(Q_2 + Q_1), \qquad S^{(b)} = -\tfrac{3}{2}(Q_2 - Q_1) \tag{4.4}$$

so that in a generic reference frame $(\hat{e}_x, \hat{e}_y, \hat{e}_z)$ has the form $(i,j = x, y, z)$

$$Q_{ij} = Q_{ij}^{(u)} + Q_{ij}^{(b)} \tag{4.5}$$

$$= S^{(u)}(n_i n_j - \tfrac{1}{3}\delta ij) + \tfrac{1}{3}S^{(b)}(n_i' n_j' - n_i'' n_j'') \tag{4.6}$$

with $n_i = \hat{n} \cdot \hat{e}_i$, $n_i' = \hat{n}' \cdot \hat{e}_i$, $n_i'' = \hat{n}'' \cdot \hat{e}_i$.

The description of the LC configuration by means of the scalar-order parameters $S^{(u)}$ (uniaxial) and $S^{(b)}$ (biaxial) and the nematic directors $\hat{n}$ and $\hat{n}'$ (note that $\hat{n} \cdot \hat{n}' = 0$ and $\hat{n} \times \hat{n}' = \hat{n}''$) is sometime preferred to the description by means of the tensor-order parameter $\mathbf{Q}$ for several reasons. First of all the scalar-order parameter only contains information about the degree of order while the nematic directors only contain information about the local LC orientation, while all such information is mixed together in the tensor-order parameter. Moreover, in almost all cases of interest to us (uniaxial nematic mesophase) only one scalar parameter $(S^{(u)})$ and one nematic director $(\hat{n})$ are required. Finally $S^{(u)}$ and $\hat{n}$ have a more intuitive physical derivation.

The 'ordinary' nematic configuration is uniaxial (N_u). This means that it has a high degree of symmetry: it has continuous rotational symmetry around nematic director $\hat{n}$, is nonchiral (symmetric with respect to spatial inversion) and nonpolar ($\hat{n}$ and $-\hat{n}$ are indistinguishable). Note that even 'polar' molecules may behave as nonpolar if any small volume contains just as many $\hat{l}$ and $-\hat{l}$ LC molecules. Here we restrict our attention to such a configuration, while nematic configurations having lower symmetry, i.e. cholesteric or chiral (N_u^*), biaxial (N_b) and chiral biaxial (N_b^*), are discussed in the literature [92–94].

The degree of alignment of the molecules can be described by means of a distribution function $f(\vartheta_l, \varphi_l)\, d\Omega_l$ giving the probability of finding molecules with molecular director $\hat{l}$ in a small solid angle $d\Omega_l = \sin\vartheta_l\, d\vartheta_l\, d\varphi_l$ around the direction (ϑ_l, φ_l). For an LC in the nematic mesophase the distribution function must be independent on φ_l (i.e. $f(\vartheta_l, \varphi_l) = f(\vartheta_l)$) since there is a complete cylindrical symmetry around $\hat{n}$ and must be symmetric with respect to $\vartheta_l = \pi/2$ (i.e. $f(\pi - \vartheta_l) = f(\vartheta_l)$) since $\hat{n}$ and $-\hat{n}$ are equivalent. We can therefore write $f = f(\cos \vartheta_l)$: the cosine of the angle between molecular and nematic director $\cos \vartheta_l = \hat{n} \cdot \hat{l}$ is a measure of the molecular alignment and the distribution function $f(\cos \vartheta_l) = f(\hat{n} \cdot \hat{l})$ describes the thermal fluctuations of the molecules around the nematic director.

Developing this function in series of Legendre polynomials P_n we obtain [5, 7]

$$f(\hat{n} \cdot \hat{l}) = \sum_0^\infty (2n + \tfrac{1}{2}) S_n P_{2n}(\hat{n} \cdot \hat{l}), \tag{4.7}$$

where the coefficients S_n are given by

$$S_n = \int_{-1}^{+1} P_{2n}(\hat{n} \cdot \hat{l}) f(\hat{n} \cdot \hat{l}) \, \mathrm{d}(\hat{n} \cdot \hat{l}) = \langle P_{2n}(\hat{n} \cdot \hat{l}) \rangle. \tag{4.8}$$

The first coefficients are

$$S_0 = \langle P_0(\hat{n} \cdot \hat{l}) \rangle = 1 \tag{4.9}$$

$$S_1 = \langle P_2(\hat{n} \cdot \hat{l}) \rangle = \tfrac{1}{2} \langle (3(\hat{n} \cdot \hat{l})^2 - 1) \rangle. \tag{4.10}$$

$$S_2 = \langle P_4(\hat{n} \cdot \hat{l}) \rangle = \tfrac{1}{8} \langle (35(\hat{n} \cdot \hat{l})^4 - 30(\hat{n} \cdot \hat{l})^2 + 3) \rangle. \tag{4.11}$$

The dimensionless quantity S_1 can be chosen as the molecular order parameter of the LC in the nematic mesophase: it is a quantity which vanishes, for symmetry reasons, in the isotropic phase while taking into account the lower symmetry (or equivalently, more order) in the nematic mesophase. Obviously its maximum value, $S_1 = 1$, corresponds to having all LC molecules perfectly aligned to each other: i.e. there is no thermal fluctuation so that $\hat{n} = \hat{l}$. It can be shown that the molecular order parameter S_1 defined by equation (4.11) is the same as the uniaxial scalar order parameter $S^{(u)}$ defined in equation (4.4). In the following we will use the symbol S.

The molecular-order parameter S is a function of the temperature T. A useful relation is

$$S(T) = S_0(1 - T/T^\dagger)^f \tag{4.12}$$

where S_0, f and $T^\dagger$ (a temperature slightly above the nematic–isotropic transition temperature T_{NI}) must be experimentally determined, has been proposed as an empirical relation and is derived [95] from a semi-empirical approach based on the Landau–de Gennes treatment. A slightly different expression, relating the order parameter to the 'reduced temperature', is

$$\tilde{T} = TV^2/(T_{NI} V_{NI}^2), \tag{4.13}$$

where V and T are the actual values of the LC volume and temperature, while V_{NI} and T_{NI} are the LC volume and temperature at which the N–I phase transition occurs, and has been derived [6] as a universal function and approximated by the analytic expression

$$S(\tilde{T}) = (1 - 0.98\tilde{T})^{0.22}. \qquad (4.14)$$

4.3.1.2 The free energy density

The director field $\hat{n}(r)$ describes the nematic director configuration. In a chosen reference frame we have $\hat{n} \equiv (\sin \vartheta \cos \varphi, \sin \vartheta \sin \varphi, \cos \vartheta)$ so that the director field is determined by the functions $\vartheta(x, y, z)$ and $\varphi(x, y, z)$. The problem of obtaining the nematic director configuration is generally very complex. It is physically determined by the interaction of the LC molecules with each other, with the limiting surfaces and with external (gravitational, electromagnetic) fields. All these effects are taken into account by means of the free energy.

$$F = \int_{\text{volume}} (\mathcal{F}_e + \mathcal{F}_{mf} + \mathcal{F}_{ef})\, dv + \int_{\text{surface}} \mathcal{F}_{si}\, ds \qquad (4.15)$$

where $\mathcal{F}_e$ is the elastic free energy density, due to 'bulk' elastic interactions, the subsequent two terms are due to magnetic ($\mathcal{F}_{mf}$) and electric ($\mathcal{F}_{ef}$) fields, and the last term ($\mathcal{F}_{si}$) is due to interaction (anchoring) of the LC molecules with boundary surfaces.

For sufficiently smooth variations of the tensor-order parameter, the elastic free energy density can be expanded in powers of the spatial derivative of the tensor-order parameter. Taking into account only terms that are linear or quadratic in the first-order derivatives and terms that are linear in the second-order derivatives, for a uniaxial nematic LC $\mathcal{F}_e$ is written as [92]

$$
\begin{aligned}
\mathcal{F}_e = &\ K_0 \hat{n} \cdot (\nabla \times \hat{n}) + K_1 (\nabla \cdot \hat{n})^2 + K_2 [\hat{n} \cdot (\nabla \times \hat{n})]^2 \\
&+ K_3 [\hat{n} \times (\nabla \times \hat{n})]^2 + K_4 (\hat{n} \cdot \nabla S)^2 + K_5 (\nabla S)^2 \\
&+ K_6 [\hat{n} \times (\nabla \times \hat{n})] \cdot (\nabla S) + K_7 (\nabla \cdot \hat{n})(\hat{n} \cdot \nabla S) \\
&+ \nabla \cdot \{K_1' [(\hat{n} \cdot \nabla)\hat{n} - \hat{n}(\nabla \cdot \hat{n})]\} + \nabla \cdot \{K_2' \nabla S\} \\
&+ \nabla \cdot \{K_3' \hat{n}(\hat{n} \cdot \nabla S)\} + \nabla \cdot \{K_4' [(\hat{n} \cdot \nabla)\hat{n} + \hat{n}(\nabla \cdot \hat{n})]\}
\end{aligned}
$$

and contains eight ($K_0, \ldots, K_7$) 'bulk' and four ($K_1', \ldots, K_4'$) surface elastic constants. The last four terms are called surface terms since, due to the presence of a divergence, the corresponding volume integrals in the free energy equation (4.15) can be transformed, applying Gauss's theorem, into surface integrals.

The first term, containing the elastic constant K_0, is nonsymmetric for spatial inversion and therefore cannot appear in ordinary (non-chiral) nematic mesophases. The fifth term, containing the elastic constant K_4, is nonsymmetric for $(\hat{n} \leftrightarrow -\hat{n})$ substitution and therefore is forbidden in nonpolar nematic LCs.

Moreover, in supramicron droplets far from the N–I transition temperature, uniform temperature and scalar order parameter are assumed, so that terms containing ∇S disappear and the elastic free energy density can be written as three volume terms and two surface terms:

$$\mathcal{F}_e = K_1(\nabla \cdot \hat{n})^2 + K_2[\hat{n} \cdot (\nabla \times \hat{n})]^2 + K_3[\hat{n} \times (\nabla \times \hat{n})]^2$$
$$+ \nabla \cdot \{K_1'[(\hat{n} \cdot \nabla)\hat{n} - \hat{n}(\nabla \cdot \hat{n})]\} + \nabla \cdot \{K_4'[(\hat{n} \cdot \nabla)\hat{n} + \hat{n}(\nabla \cdot \hat{n})]\}.$$

It can be shown [94] that it is equivalent to the classical notation introduced by Frank and by Nehring and Saupe [96]

$$\mathcal{F}_e = \tfrac{1}{2}K_{11}(\nabla \cdot \hat{n})^2 + \tfrac{1}{2}K_{22}[\hat{n} \cdot (\nabla \times \hat{n})]^2 + \tfrac{1}{2}K_{33}[\hat{n} \times (\nabla \times \hat{n})]^2$$
$$- \nabla \cdot \{K_{13}\hat{n}(\nabla \cdot \hat{n})\}$$
$$- \nabla \cdot \{(K_{22} + K_{24})[\hat{n}(\nabla \cdot \hat{n}) + \hat{n} \times (\nabla \times \hat{n})]\}$$

where the splay (K_{11}), twist (K_{22}), bend (K_{33}), splay–bend (K_{13}) and saddle–splay (K_{24}) Frank elastic constants are used.

The surface terms are usually neglected in determining the configuration of a plane slab of nematic LC, but significantly contribute to free energy density of LC in confined volumes where there is a high surface-to-volume ratio. However, these terms mainly contribute to determine the stability of the nematic director configuration, so they are sometimes ignored even in calculating the nematic director field inside droplets.

The effect of the interaction (anchoring) of LC molecules with surrounding polymer at the droplet surface is taken into account by means of the 'anchoring free energy density' term [97]

$$\mathcal{F}_{si} = \tfrac{1}{2}(W_\varphi \sin^2 \varphi_s + W_\vartheta \cos^2 \vartheta_s) \sin^2 \vartheta_s, \qquad (4.16)$$

where the two constants W_φ and W_ϑ are the azimuthal and polar anchoring strength, φ_s and ϑ_s are the angles determining the orientation of the nematic director $\hat{n}$ with respect to the surface director $\hat{s}$ and the preferred direction $\hat{n}_s$. The azimuthal term is meaningless for homeotropic anchoring ($\hat{n}_s = \hat{s}$), while the whole anchoring energy density can be neglected for both strong ($\hat{n} = \hat{n}_s$ and hence $\vartheta_s = 0$) or negligible ($W_\varphi = W_\vartheta = 0$) anchoring.

Finally, the free energy densities due to external electric and magnetic fields are

$$\mathcal{F}_{ef} = -\tfrac{1}{2}E \cdot D, \qquad \mathcal{F}_{mf} = -\tfrac{1}{2}B \cdot H, \qquad (4.17)$$

which for uniaxial nematic LCs can be written

$$\mathcal{F}_{ef} = -\frac{\varepsilon_0}{2}\Delta\varepsilon(E \cdot \hat{n})^2, \qquad \mathcal{F}_{mf} = -\frac{1}{2\mu_0}\Delta\chi(B \cdot \hat{n})^2, \qquad (4.18)$$

where ε_0 and μ_0 are electrical and magnetic permittivity in a vacuum and $\Delta\varepsilon = \varepsilon_\| - \varepsilon_\perp$ and $\Delta\chi = \chi_\| - \chi_\perp$ are dielectric and diamagnetic aniso-tropies. It must be noted that local and not externally applied fields must be considered.

4.3.2   Light scattering in PDLCs

Light passing through a PDLC film is affected by the presence of LC droplets acting as 'scattering objects'. Here we will first consider the effect of a single scatterer and then the whole PDLC film acting as a slab of scatterers.

4.3.2.1   Single scatterer

Let us consider a single scattering object (i.e. a PDLC droplet) O in a homo-geneous, isotropic, non-absorbing medium (the polymer) characterized by refractive index n_m, and an incoming monochromatic linearly polarized plane wave (the incident beam) travelling in the direction defined by the wavevector $k_m^{(i)}$ whose magnitude is $k_m = 2\pi n_m/\lambda_0$ with λ_0 being the wave-length in a vacuum. The fundamental question in a light scattering problem [99] is to determine the electromagnetic field at an arbitrary point P in the surrounding medium. This can be accomplished by taking into account the incoming electromagnetic field and the scattered one, due to the super-position of the radiation generated by each dipole (LC molecule) inside the scattering object.

It is virtually impossible to obtain an answer to this question in such a general form, even assuming a known dipole distribution inside the scat-tering object, since each dipole is excited by both the incoming and the scattered radiation. Therefore some approximation, depending on the specific problem (fundamentally size and refractive index mismatch of the scattering object), has to be introduced, but let us first develop some general considerations [98].

- Raman scattering does not occur in non-dye-doped LCs, so we assume that no wavelength change occurs between incoming and scattered light: $\lambda_m^{(s)} = \lambda_m^{(i)} = \lambda_m = \lambda_0/n_m$.
- If we are just interested in the *far-field* (i.e. $k_m r_{OP} \gg 1$, $r_{OP} = |r_{OP}|$ being the distance of observation point P from scattering object O) the scattered electromagnetic field in P can be locally expressed as a spherical wave with its source in O. Let us call $k_m^{(s)}$ its wavevector whose magnitude is the same as the incoming one ($|k_m^{(s)}| = |k_m^{(i)}| = k_m = 2\pi/\lambda_m$ where the subscript 'm' is used to indicate that we are dealing with quantities concerning the medium surrounding the scattering object).

Looking at figure 4.5, the scattering direction $\hat{e}_k^{(s)} = k_m^{(s)}/k_m$ and the incident direction $\hat{e}_k^{(i)} = k_m^{(i)}/k_m$ define the scattering angle ϑ (the angle between them)

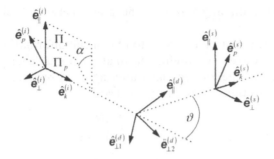

Figure 4.5. Angles and reference frames defined in describing light scattering.

and, for any off-axis ($\hat{e}_m^{(s)} \neq \pm \hat{e}_m^{(i)}$) point P, the scattering plane Π_s. The polarization direction $\hat{e}_p^{(i)}$ and the incident direction define the polarization plane Π_p and the polarization angle α (the angle between Π_p and Π_s).

The incident electromagnetic wave ($\Re[\ ]$ means real part)

$$E^{(i)}(r, t) = \Re[\exp(ik_m^{(i)} \cdot r) E_0^{(i)} \hat{e}_p^{(i)} \exp(i\omega t)] \qquad (4.19)$$

is more conveniently expressed using the orthonormal vectors $(\hat{e}_\perp^{(i)}, \hat{e}_\parallel^{(i)}, \hat{e}_k^{(i)})$ with $\hat{e}_\parallel^{(i)}$ lying in the scattering plane:

$$E^{(i)} = \Re[\exp(ik_m^{(i)} \cdot r)(E_\parallel^{(i)} \hat{e}_\parallel^{(i)} + E_\perp^{(i)} \hat{e}_\perp^{(i)}) \exp(i\omega t)], \qquad (4.20)$$

where

$$E_\parallel^{(i)} = E_0^{(i)} \hat{e}_p^{(i)} \cdot \hat{e}_\parallel^{(i)} = E_0^{(i)} \cos \alpha, \qquad (4.21)$$

$$E_\perp^{(i)} = E_0^{(i)} \hat{e}_p^{(i)} \cdot \hat{e}_\perp^{(i)} = E_0^{(i)} \sin \alpha. \qquad (4.22)$$

The far-field scattered electromagnetic wave, expressed using the orthonormal vectors $(\hat{e}_\perp^{(s)}, \hat{e}_\parallel^{(s)}, \hat{e}^{(s)})$ with $\hat{e}_\parallel^{(s)}$ lying in the scattering plane, is

$$E^{(s)} = \Re\left[\frac{\exp(ik_m r_{\mathrm{OP}})}{-ik_m r_{\mathrm{OP}}} (E_\perp^{(s)} \hat{e}_\perp^{(s)} + E_\parallel^{(s)} \hat{e}_\parallel^{(s)}) \exp(i\omega t) \right] \qquad (4.23)$$

or

$$E^{(s)} = \Re\left[\frac{\exp(ik_m r_{\mathrm{OP}})}{-ik_m r_{\mathrm{OP}}} (E_0^{(i)} \mathbf{S} \cdot \hat{e}_p^{(i)}) \exp(i\omega t) \right], \qquad (4.24)$$

so that the problem is to determine the four (complex) elements of the scattering matrix $\mathbf{S}$ connecting incident and scattered electromagnetic waves:

$$\begin{bmatrix} E_\parallel^{(s)} \\ E_\perp^{(s)} \end{bmatrix} = \begin{bmatrix} S_2 & S_3 \\ S_4 & S_1 \end{bmatrix} \cdot \begin{bmatrix} E_\parallel^{(i)} \\ E_\perp^{(i)} \end{bmatrix}. \qquad (4.25)$$

The scattering coefficients $S_i(\vartheta, \alpha)$ are generally functions of both the scattering angle ϑ (i.e. the angle between scattering and incident directions) and the polarization angle α (i.e. the angle between polarization and scattering planes).

The scattering matrix allows us to obtain full information on scattered electromagnetic wave, but useful information about energy distribution can be obtained by the 'differential scattering cross sections' and the 'total scattering cross sections' respectively defined by

$$\frac{d\sigma}{d\Omega} = \frac{|\boldsymbol{E}^{(s)}|^2}{|\boldsymbol{E}^{(i)}|^2} r_{OP}^2, \qquad \sigma = \int_0^\pi \int_0^{2\pi} \frac{d\sigma}{d\Omega} \sin\vartheta \, d\vartheta \, d\alpha. \tag{4.26}$$

Taking into account equations (4.24) and (4.19), we obtain

$$\frac{d\sigma}{d\Omega} = \frac{\pi}{k_m^2} |\boldsymbol{S} \cdot \hat{\boldsymbol{e}}_p^{(i)}|^2 \tag{4.27}$$

or, explicitly,

$$\frac{d\sigma}{d\Omega} = \frac{\pi}{k_m^2} (|S_2 \cos\alpha + S_3 \sin\alpha|^2 + |S_4 \cos\alpha + S_1 \sin\alpha|^2). \tag{4.28}$$

With a little more computation, and using the so-called 'optical theorem' [99, 100] it is possible to compute the total scattering cross-section just knowing the 'forward' (i.e. for $\vartheta = 0$) scattering coefficients

$$\sigma = \frac{4\pi^2}{k_m^2} \Re[\hat{\boldsymbol{e}}_p^{(i)} \cdot \boldsymbol{S}_{\vartheta=0} \cdot \hat{\boldsymbol{e}}_p^{(i)}] \tag{4.29}$$

or, explicitly,

$$\sigma = \frac{4\pi^2}{k_m^2} \Re[S_{1,\vartheta=0} \sin^2\alpha + S_{2,\vartheta=0} \cos^2\alpha + (S_{3,\vartheta=0} + S_{4,\vartheta=0}) \sin\alpha \cos\alpha]. \tag{4.30}$$

This apparent paradox has an easy physical interpretation. Transmitted light is the result of interference of incident and forward scattered ($\vartheta = 0$) electromagnetic waves and, therefore, is fully determined by $\boldsymbol{S}_{\vartheta=0}$. Total scattered energy is determined by σ but the conservation of energy (we are in a nonabsorbing medium) requires that it must be equal to incoming minus transmitted energy, so that, just like transmitted light, must be fully determined by $\boldsymbol{S}_{\vartheta=0}$.

Scattering matrix or scattering cross-sections can be used to determine the effect of a PDLC film, i.e. a slab filled with scattering objects.

Rigorous solution of the scattering problem can be obtained (Mie theory) [98] for a homogeneous isotropic sphere (radius R_s, refractive index n_s) in an infinite homogeneous isotropic medium (refractive index n_m). Such a case is not of interest for application to PDLCs (non-spherical, inhomogeneous, anisotropic LC droplets) but it is interesting to analyse the role of some parameters in obtaining simplified solutions [98]. The two fundamental parameters are the normalized refraction index mismatch $|\delta n/n_m| = |(n_s - n_m)/n_m|$ and the normalized scatterer size $\tilde{D} = 2R_s k_m$. Note that refractive indices are real numbers since we assume that both scattering object (non-dye-doped PDLC droplets) and the surrounding medium (polymer) are non-absorbing. A third parameter, related to the first two, is the maximum phase shift (i.e. the change of phase of a ray passing through the sphere along a diameter): $|\Delta\Phi_{max}| = (2\pi/\lambda_0)2R_s|n_s - n_m| = \tilde{D}|\delta n/n_m|$. All these dimensionless parameters can assume any value in the range $(0,\infty)$ and with respect to their value we have different simplified situations: optically small ($\tilde{D} \ll 1$) or large ($\tilde{D} \gg 1$), optically soft ($|\delta n/n_m| \ll 1$) or hard ($|\delta n/n_m| \gg 1$), optically thin ($|\Delta\Phi_{max}| \ll 1$) or thick ($|\Delta\Phi_{max}| \gg 1$) scattering objects. We must underline that here the optical thickness is referred to the alteration produced on the phase and not on the intensity of light. Any combinations of two of these conditions, without any condition on the remaining parameter, lead to situations with especially clear physical interpretation and particularly simple solutions of the scattering problem.

(i) Optically small thin scatterer (Rayleigh scattering). The external field is assumed to be homogeneous within the scattering object (small) and penetrating so fast (thin) that the static polarization is established instantaneously, independently from the refractive index mismatch value.

(ii) Optically thin soft scatterer (Rayleigh–Gans). The scatterer may be large with respect to wavelength but the refractive index mismatch is small enough (soft) to neglect reflection, and the crossing time is short enough (thin) to assume all parts of the scattering object to be excited in phase, so that scattered field may be seen as simultaneous interfering Rayleigh scattering of all elements within the scattering object. In other words, in Rayleigh–Gans approximation (RGA), analogous to the Born approximation in quantum mechanics, scattered electromagnetic far-field is assumed to be the total dipole radiation generated by all dipoles (LC molecules) in the scattering object (PDLC droplet), assuming that each dipole is excited only by the incident electromagnetic field.

(iii) Optically soft large scatterer (anomalous diffraction). The refractive index mismatch is small enough (soft) to neglect reflection so that phase-only shift is taken into account, replacing the three-dimensional scattering object with a two-dimensional (its projection in a plane orthogonal to the scattering direction) phase object.

(iv) Optically large thick scatterer (large spheres). The scattering problem can be treated as a combination of a diffraction problem (dependent on $\tilde{R}$ but not on $\tilde{n}$) and a reflection/refraction problem (dependent on $\tilde{n}$ but not on $\tilde{R}$).

(v) Optically thick hard scatterer (total reflector). The electromagnetic field within the scatterer may be neglected and the scattering field may be assumed to be produced by reflection at the scattering object boundary.

(vi) Optically hard small scatterer (optical resonance). The index mismatch is large (hard) so that reflection at scatterer boundaries cannot be neglected, so that there are some electromagnetic vibration modes which are nearly self-sustained: for some critical combinations of the parameters, a vibration of large amplitude can be sustained by an incident wave of relatively small amplitude, but the same is not true for slightly different values of the parameters.

To achieve a good transparency of the PDLC film in the ON-state, the refractive index of the polymer must be equal or very similar to the refractive index of the ordered (ON-state) droplets. This means that we generally deal with optically soft scatterers, so that only the first four cases, mainly RGA and anomalous diffraction approach (ADA), are relevant in studying PDLCs. An interesting situation is obtained when both approximations are valid (optically thin–soft–large scatterer), i.e. if refractive index mismatch is small enough to produce a small phase shift even if the scattering object is large with respect to wavelength.

4.3.2.2   A slab of scatterers

Let us consider a slab of thickness h (i.e. a region defined by $-\infty \leq x \leq +\infty$, $-\infty \leq y \leq +\infty$, $-h/2 \leq z \leq +h/2$) containing a more or less uniform distribution of scattering objects and a plane wave impinging on first boundary surface $z = -h/2$. We assume that there is no difference in the optical properties of the continuous medium filling the space between scattering objects and the continuous medium filling the space outside the slab, since the boundary effects (reflection and refraction) at $z = -h/2$ and $z = +h/2$ can be easily added.

The electric field at any point $P \equiv (x, y, z)$, i.e. inside ($|z| < h/2$) and outside ($|z| > h/2$) the slab, is given by the sum of the incident field and the fields scattered by each scatterer. As a consequence, with respect to the single scatterer situation, two more effects must be taken into account:

• inside the slab: the electric field experienced by each scattering object is influenced by the presence of the other scattering objects (multiple scattering effect);

• outside the slab: the light scattered by different particles interfere with each other (structure factor effect).

A rigorous treatment of multiple scattering is an extremely complicated problem, even for a distribution of isotropic spheres [101], and will not be considered here. The multiple scattering effect can be neglected if the particles are not too densely packed and the slab is thin enough.

Let us now develop some considerations about the structure factor. If we are interested in determining the light scattered by the whole slab it is convenient to define the effective differential cross section

$$\frac{\mathrm{d}\sigma_s}{\mathrm{d}\Omega} = \frac{1}{N_s} \frac{|E^{(s)}_{\text{tot}}|^2}{|E^{(i)}|^2} r^2, \tag{4.31}$$

where N_s is the number of scattering objects involved and $E^{(s)}_{\text{tot}}$ is the total scattered electric field at point P defined by position vector $\boldsymbol{r}$. In the far-field approximation

$$\boldsymbol{E}^{(s)}_{\text{tot}} = \sum_j \boldsymbol{E}^{(s)}_j \exp(-\mathrm{i}\Delta\boldsymbol{k}\cdot\boldsymbol{r}_j), \tag{4.32}$$

where $\boldsymbol{E}^{(s)}_j$, the electric field scattered by the scattering object placed at $\boldsymbol{r}_j$, is given by equation (4.24). The phase term, where $\Delta\boldsymbol{k} = \boldsymbol{k}^{(s)} - \boldsymbol{k}^{(i)}$, is required because in equation (4.24) we considered a scattering object placed at the origin of reference frame.

Substituting equation (4.32) into equation (4.31) gives

$$\frac{\mathrm{d}\sigma_s}{\mathrm{d}\Omega} = \frac{1}{N_s} \sum_j \frac{|E^{(s)}_j|^2}{|E^{(i)}|^2} r^2 \tag{4.33}$$

$$+ \frac{1}{N_s} \sum_{i,j \neq i} \frac{\boldsymbol{E}^{(s)}_i \cdot \boldsymbol{E}^{*(s)}_j}{|E^{(i)}|^2} \exp[-\mathrm{i}\Delta\boldsymbol{k}\cdot(\boldsymbol{r}_i - \boldsymbol{r}_j)] r^2. \tag{4.34}$$

The first term is the 'slab average' of the scattering cross-section

$$\frac{1}{N_s} \sum_j \frac{|E^{(s)}_j|^2}{|E^{(i)}|^2} r^2 = \frac{1}{N_s} \sum_j \frac{\mathrm{d}\sigma_j}{\mathrm{d}\Omega} = \left\langle \frac{\mathrm{d}\sigma}{\mathrm{d}\Omega} \right\rangle_s \tag{4.35}$$

but this can also be interpreted as the scattering cross-section averaged over all possible droplet orientations and sizes. The second term, assuming that there is no correlation between droplet position and its size or orientation, is usually written as [102]

$$\frac{1}{N_s} \sum_{i,j \neq i} \frac{\boldsymbol{E}^{(s)}_i \cdot \boldsymbol{E}^{*(s)}_j}{|E^{(i)}|^2} \exp[-\mathrm{i}\Delta\boldsymbol{k}\cdot(\boldsymbol{r}_i - \boldsymbol{r}_j)] r^2 = \frac{\mathrm{d}\sigma_{\langle j \rangle}}{\mathrm{d}\Omega} G_s(\Delta\boldsymbol{k}) \tag{4.36}$$

where $\sigma_{\langle j \rangle}$ is the scattering cross-section corresponding to the averaged value of the scattered field and

$$G_s(\Delta\boldsymbol{k}) = \frac{1}{N_d} \sum_j \exp[-\mathrm{i}\Delta\boldsymbol{k}\cdot(\boldsymbol{r}_i - \boldsymbol{r}_j)] \tag{4.37}$$

is a 'correlation function'.

The effective differential cross-section can thus be written

$$\frac{d\sigma_s}{d\Omega} = \left\langle \frac{d\sigma}{d\Omega} \right\rangle_s + \frac{d\sigma_{(j)}}{d\Omega} G_s(\Delta k) \tag{4.38}$$

or

$$\frac{d\sigma_s}{d\Omega} = \left\langle \frac{d\sigma}{d\Omega} \right\rangle_s F_s(\Delta k) \tag{4.39}$$

if we define the structure factor

$$F_s(\Delta k) = 1 + G_s(\Delta k) \frac{d\sigma_{(j)}}{d\Omega} \left(\left\langle \frac{d\sigma}{d\Omega} \right\rangle_s \right)^{-1}. \tag{4.40}$$

Structure factor is related to spatial distribution of scattering objects within the slab.

Finally the intensity of the beam passing through the slab is given by the classical law

$$I(z) = I_0 \exp(-\rho_n \sigma_s z), \tag{4.41}$$

where I_0 is beam intensity at $z = 0$, ρ_n is the number of scattering objects per unit volume and σ_s is the effective scattering cross-section.

4.4   PDLC electro-optical behaviour

The theory delineated in the previous section should allow the determination of the droplet configuration and the scattering cross-section of an LC droplet and consequently the electro-optical behaviour of a PDLC film. Unfortunately the problem is so complex that no general solution can be found. Two approaches have been followed, based on computer simulation or on some physically reasonable simplifications. In most cases numerical computation must be performed. Computer simulations may give better results in specific situations, but we think that approximations lead to a better understanding of the phenomena. For such reasons this section will be mainly devoted to examining approximate solutions.

4.4.1   Droplet configurations

The most common approximation made in studying nematic LC distribution inside PDLC droplets is to assume a droplet having ellipsoidal shape. The major axis defines a preferred direction, usually called the zero-field droplet director, $\hat{n}_d^*$. An even simpler assumption is a spherical droplet characterized by a radius R_d and a preferred direction $\hat{n}_d^*$. The presence of an external field changes the nematic director distribution, but it is always possible to define a droplet director $\hat{n}_d$ as the average direction of the nematic director $\hat{n}$ inside the droplet. Obviously $\hat{n}_d|_{\text{zero field}} = \hat{n}_d^*$.

Another usual assumption is the so-called one elastic constant approximation ($K_{11} = K_{22} = K_{33} = K$). This approximation is sometimes relaxed, defining the twist and bend relative elastic constants

$$\kappa_T = K_{22}/K_{11}, \qquad \kappa_B = K_{33}/K_{11}, \qquad (4.42)$$

and assuming $\kappa_T = \kappa_B = \kappa$.

4.4.1.1 *Analytical/numerical approach*

Using cylindrical coordinates the nematic director field is described by the function $\vartheta_n(\rho, \varphi, z)$ giving the angle between the nematic director $\hat{n}(\rho, \varphi, z)$ and droplet director $\hat{n}_d$. It has been shown [103] that under the assumptions of (i) strong anchoring, (ii) one elastic constant and (iii) spherical droplet, minimization of free energy leads to the nonlinear partial differential equation

$$\nabla^2 \vartheta_n - \left(\frac{1}{\xi^2} + \frac{1}{\rho^2}\right) \cos \vartheta_n \sin \vartheta_n = 0, \qquad (4.43)$$

where ξ is the field correlation length and is given by

$$\xi = \xi_m = \left(\frac{\mu_0 K}{\Delta \chi}\right)^{1/2} \frac{1}{B} \quad \text{or} \quad \xi = \xi_e = \left(\frac{K}{\varepsilon_0 \Delta \varepsilon}\right)^{1/2} \frac{1}{E} \qquad (4.44)$$

in the presence of a magnetic or electric field, respectively.

Solving numerically the differential equation with different boundary conditions, corresponding to different anchoring conditions, and different field strength, measured by the ratio $\tilde{F} = R_d/\xi$, allows us to obtain typical droplet configurations. Zero field ($\tilde{F} = 0$) and strong anchoring leads to bipolar (figure 4.6(a)) or radial (figure 4.6(b)) droplet configurations, depending on the type of anchoring (planar or homeotropic, respectively) [74, 103].

Removing the one-constant approximations, other configurations are found. Most common are toroidal (figure 4.6(c)) found assuming planar surface anchoring and $K_{33} \ll K_{11}$ and axial (figure 4.6(d)) found assuming homeotropic surface anchoring and $K_{33} \gg K_{11}$.

All these four basic configurations (figure 4.6) can be represented by approximate analytical expressions (table 4.2) giving molecular director angles (ϑ_n and φ_n) as a function of the position of the LC molecule inside the droplet. The origin of the spherical coordinate system ($\tilde{\rho}$, ϑ, φ) is chosen at the centre of the spherical droplet, the radial coordinate is normalized ($\tilde{\rho} = \rho/R_d$) and the $\vartheta = 0$ axis is aligned to the droplet director.

The 'uniform' configuration is sometimes used since it is reasonably similar to bipolar and axial ones but leads to more simple calculations.

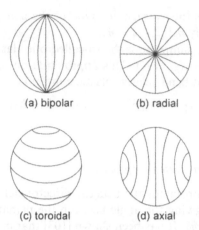

(a) bipolar   (b) radial

(c) toroidal   (d) axial

Figure 4.6. Common nematic director configurations: (a) bipolar, (b) radial, (c) toroidal, (d) axial.

Moreover it is also a good approximation for the nematic director distribution under high electric field.

More complex configurations are possible: for example it has been shown [104] that a twisted configuration (figure 4.7) occurs for large values ($K_{11} \geq K_{22} + 0.431 K_{33}$) of the splay elastic constant. Discussion on nematic droplet configurations and the transition between different configurations can be found in the literature [48].

4.4.1.2 Computer simulation approach

A different approach to determine nematic droplet configuration is obtained using Monte Carlo computer simulation. A Lebwohl–Lasher model [105,

Table 4.2. Nematic director distributions for basic droplet configurations in normalized ($\tilde{\rho} = \rho/R_d \in [0, 1]$) spherical coordinates with origin at droplet centre.

Droplet configuration	$\vartheta_n (\tilde{\rho}, \vartheta, \varphi)$	$\varphi_n (\tilde{\rho}, \vartheta, \varphi)$
Bipolar	$\left(\vartheta - \dfrac{\pi}{2}\right)\tilde{\rho}$	φ
Radial	ϑ	φ
Toroidal	$\frac{1}{2}\pi$	$\varphi + \frac{1}{2}\pi$
Axial	$\vartheta + \arctan\left[\dfrac{1 + \tilde{\rho}^2}{1 - \tilde{\rho}^2}\dfrac{1}{\tan\vartheta}\right] + \dfrac{\pi}{2}$	φ
Uniform	0	φ

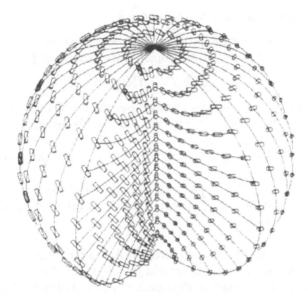

Figure 4.7. Nematic director field for relative elastic constant values $\kappa_T = 0.6$ and $\kappa_B = 0.5$. From R D Williams 1986 *J. Phys. A: Math. Gen.* **19** 3211.

106] assuming discretized positions and continuously varying orientations is followed. The droplet is schematized by a sphere containing a cubic lattice of 'particles' each characterized by unit vectors ($\hat{\boldsymbol{u}}_i$, $i \in \mathbb{S}$). The term particle may refer to a single LC molecule ($\hat{\boldsymbol{u}} = \hat{\boldsymbol{l}}$) or a group of LC molecules ($\hat{\boldsymbol{u}} = \hat{\boldsymbol{n}}$). Particle interactions are described by means of a nearest-neighbour potential containing three terms

$$U_N = -\sum_{i,j \in \mathbb{S}} \varepsilon_{ij} P_2(\hat{\boldsymbol{u}}_i \cdot \hat{\boldsymbol{u}}_j) - \sum_{i \in \mathbb{S}, j \in \mathbb{G}} \varepsilon_{ij} J P_2(\hat{\boldsymbol{u}}_i \cdot \hat{\boldsymbol{u}}_j) - \varepsilon \xi_F \sum_{i \in \mathbb{S}} P_2(\hat{\boldsymbol{u}}_i \cdot \hat{\boldsymbol{u}}_F),$$

$$(4.45)$$

where P_2 is the Legendre polynomial of order 2.

The first term takes into account the elastic interaction between bulk particles by means of ε_{ij}, a positive constant ε for adjacent ($\hat{\boldsymbol{u}}_i$, $\hat{\boldsymbol{u}}_j$) particles and zero otherwise. The second term, containing parameter J, takes into account the surface anchoring by means of 'ghost' particles [107] having fixed orientation [108] ($\hat{\boldsymbol{u}}_i$, $i \in \mathbb{G}$) placed in a layer outside the spherical surface. The last term takes into account, by means of parameter ξ_F, the effect of an electric or magnetic field in the direction $\hat{\boldsymbol{u}}_F$.

Such a computer simulation approach has been followed to investigate droplet configurations with radial [108, 109], toroidal [110] and bipolar [111] surface conditions. It has recently been extended to study dynamic behaviour of PDLCs [112].

4.4.1.3   Multiple-order parameters approach

A useful parameter describing the more or less good alignment of the LC molecules inside a droplet is the 'droplet order parameter' [29] defined as

$$S_d = \langle P_2(\hat{\boldsymbol{n}}_d \cdot \hat{\boldsymbol{n}}) \rangle = \tfrac{1}{2} \langle (3(\hat{\boldsymbol{n}}_d \cdot \hat{\boldsymbol{n}})^2 - 1) \rangle_d, \qquad (4.46)$$

where the symbol $\langle \rangle_d$ means an average over the droplet volume.

Comparing this definition with equation (4.10) defining the molecular order parameter, it is clear that a droplet with $S_d = 0$ roughly behaves as an isotropic object while a droplet with $S_d = 1$ behaves as a perfectly uniaxial body.

Exercise 1. Compute the droplet order parameter S_d for the basic droplet configurations, using the approximate analytical expressions of the nematic director fields shown in table 4.2.

Usually the zero-field droplet directors $\hat{\boldsymbol{n}}_d^*$ are randomly oriented within the PDLC film, but special preparation techniques or the application of an external field induces a partial alignment of LC molecules and therefore of the droplet directors $\hat{\boldsymbol{n}}_d$. If we define the film director $\hat{\boldsymbol{n}}_f$ as the privileged direction induced by an external electric $(\boldsymbol{E} = E\hat{\boldsymbol{n}}_f)$ or magnetic $(\boldsymbol{H} = H\hat{\boldsymbol{n}}_f)$ field, it is possible to define the film order parameter:

$$S_f = \langle P_2(\hat{\boldsymbol{n}}_f \cdot \hat{\boldsymbol{n}}_d) \rangle = \tfrac{1}{2} \langle (3(\hat{\boldsymbol{n}}_f \cdot \hat{\boldsymbol{n}}_d)^2 - 1) \rangle_d. \qquad (4.47)$$

We can assume that the zero-field film order parameter is zero, $S_f^* = 0$, unless special manufacturing techniques (e.g. electric field applied during polimerization [113, 114], or film stretching after polymerization [59, 86] or use of anisotropic polymers [81]) are used.

Order parameters can be used to express, under some reasonable hypotheses, the free energy density

$$\mathcal{F} = -\tfrac{1}{3} K_d S S_d P_2(\hat{\boldsymbol{n}}_d \cdot \hat{\boldsymbol{n}}_d^*) \qquad (4.48)$$

$$- \tfrac{1}{3} \mu_0 \Delta\chi H^2 S S_d P_2(\hat{\boldsymbol{n}}_d \cdot \hat{\boldsymbol{n}}_f) \qquad (4.49)$$

$$- \tfrac{1}{3} g\varepsilon_0 \Delta\varepsilon E^2 S S_d P_2(\hat{\boldsymbol{n}}_d \cdot \hat{\boldsymbol{n}}_f). \qquad (4.50)$$

Let us now discuss the three terms appearing in this expression. It has been shown [115] that under the one elastic constant approximation and assuming droplets to be axially symmetric ellipsoids, the volume elastic term of the free energy can be written as

$$\mathcal{F}_{ve} = -\frac{1}{3} K \frac{\zeta^2}{R_e^2} S S_d P_2(\hat{\boldsymbol{n}}_d \cdot \hat{\boldsymbol{n}}_d^*), \qquad (4.51)$$

and it is reasonable [116] that, even relaxing the hypotheses on the shape of the droplets and on the elastic constants, it is possible to write equation (4.48) where K_d is a parameter depending on the LC elastic constants and on droplet size and shape. It can be considered as an 'effective elastic constant per unit surface' taking into account the restoring torque which, after external field is removed, induces relaxation of LC molecules to their zero-field orientation.

It has also been shown [115] that a similar expression, equation (4.9), can be used for the magnetic field free energy density under the assumption that the magnetic field H is unaffected by the orientation of LC molecules.

The electric field term of the free energy requires some additional consideration: parameter g is required since we cannot assume that the electric field is unaffected by LC molecules orientation. We assume that the electric field can be determined using an average LC dielectric constant (ε_{LC}) and an average PDLC film dielectric constant (ε_f). Several approximate expression have been proposed [57, 84, 115, 117, 118], all giving similar numerical results:

$$\varepsilon_f = \varepsilon_p + 3v_{LC}\varepsilon_f \frac{\varepsilon_{LC} - \varepsilon_p}{2\varepsilon_f + \varepsilon_{LC}}, \tag{4.52}$$

$$1 - v_{LC} = \frac{\varepsilon_{LC} - \varepsilon_f}{\varepsilon_{LC} + \varepsilon_p} \left(\frac{\varepsilon_p}{\varepsilon_f}\right)^{1/3}, \tag{4.53}$$

$$\varepsilon_f = \varepsilon_p + 3v_{LC}\varepsilon_p \frac{\varepsilon_{LC} - \varepsilon_p}{\varepsilon_{LC} + 2\varepsilon_p - v_{LC}(\varepsilon_{LC} - \varepsilon_p)}, \tag{4.54}$$

$$\varepsilon_f = \varepsilon_p + v_{LC}(\varepsilon_{LC} - \varepsilon_p), \tag{4.55}$$

where v_{LC} is the LC volume fraction, ε_p is the polymer dielectric constant and ε_{LC} is the average LC dielectric constant. For identical ellipsoidal droplets randomly oriented [115], but reasonably also for other shapes and distributions,

$$\varepsilon_{LC} = \varepsilon_\perp + \tfrac{1}{3}(1 + SS_dS_f)\Delta\varepsilon, \tag{4.56}$$

where $\varepsilon_\parallel$, $\varepsilon_\perp$ and $\Delta\varepsilon = \varepsilon_\parallel - \varepsilon_\perp$ are LC dielectric constants and dielectric anisotropy. The simplest expression for g is obtained using equations (4.55) and (4.56), which leads to $g = v_{LC}$.

This allows us to compute the film order parameter S_f (see figure 4.8)

$$S_f = \frac{1}{4} + \frac{3}{16}\frac{F_r^2 + 1}{F_r^2} + \frac{3}{32}\frac{(3F_r^2 + 1)(F_r^2 - 1)}{F_r^3}\ln\left|\frac{F_r + 1}{F_r - 1}\right| \tag{4.57}$$

as a function of the 'reduced field' given by

$$F_r = H\left(\mu_0 \frac{\Delta\chi}{K_d}\right)^{1/2} \tag{4.58}$$

$$F_r = E\left(v_{LC}\varepsilon_0 \frac{\Delta\varepsilon}{K_d}\right)^{1/2} \tag{4.59}$$

for a magnetic and electric field, respectively.

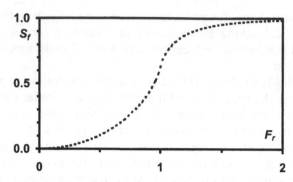

Figure 4.8. Film order parameter (S_f) versus reduced field (F_r).

4.4.2 Scattering cross section

The two approaches used in studying scattering from a single LC droplet are
RGA and ADA. The first is valid for

$$|\delta n/n_m| \ll 1, \qquad 2R_s k_m |\delta n/n_m| \ll 1, \tag{4.60}$$

while the second is valid for

$$|\delta n/n_m| \ll 1 \qquad 2R_s k_m \gg 1. \tag{4.61}$$

This is sometimes interpreted by saying that RGA is valid for small
($R_s \ll \lambda_0/(4\pi\delta n)$) while ADA is valid for large ($R_s \gg \lambda_0/(4\pi n_m)$) droplets.
However we must note that both conditions may be verified. Typical
values for refractive indices are $n_e = 1.7$, $n_o = 1.5$, $n_m = 1.52$ (n_m must be
slightly larger than n_o to get good transmittance for non-perfectly aligned
LC droplets) so that $|\delta n| < 0.18$ and $|\delta n|/n_m < 0.12$. Considering a wave-
length $\lambda_0 = 600$ nm, in the middle of a visible spectrum, we have

$$\text{RGA:} \quad R_s \ll 256 \text{ nm,} \qquad \text{ADA:} \quad R_s \gg 31 \text{ nm,} \tag{4.62}$$

so that both RGA or ADA can be reasonably applied if the droplet radius is
$R_s = 150$ nm.

4.4.2.1 Anomalous diffraction approach

The three-dimensional scattering object is replaced by a two-dimensional
phase object $A_\perp$ placed in a plane perpendicular to wavevector $k_m^{(i)}$. The
scattering matrix is given by

$$\mathbf{S} = \frac{k_m^2}{2\pi} \int_{A_\perp} [1 - \mathbf{P}(r'')] \exp(i k' \cdot r'') \, dA, \tag{4.63}$$

where the matrix $\mathbf{P}$ takes into account the phase and polarization changes
induced on the impinging beam. The matrix $\mathbf{P}$, the scattering matrix $\mathbf{S}$ and

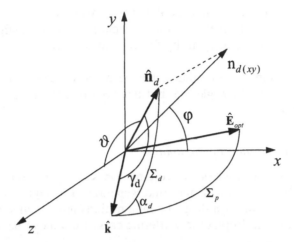

Figure 4.9. Geometrical definitions used in studying light scattering from an LC droplet.

finally the scattering cross-section σ have been determined for some simple droplet configurations.

Explicit calculations have been performed [100] for spherical droplets with radial or uniform nematic director distribution. The latter case gives simple and useful results for the scattering cross section:

$$\sigma = 2\sigma_0[\cos^2 \alpha_d\, H(s_e) + \sin^2 \alpha_d\, H(s_o)], \qquad (4.64)$$

where $\sigma_0 = \pi R_s^2$ is the geometrical cross-section,

$$H(s) = 1 - 2\,\frac{\sin(2k_m R_s s)}{2k_m R_s s} + 2\,\frac{1 - \cos(2k_m R_s s)}{(2k_m R_s s)^2} \qquad (4.65)$$

and

$$s_e = \frac{n_{de}^*}{n_m} - 1, \qquad s_o = \frac{n_{do}^*}{n_m} - 1, \qquad (4.66)$$

while n_{do}^* and n_{de}^* are the ordinary and extraordinary 'effective refractive indices' for light impinging in the direction defined by $k_m^{(i)}$. For a uniaxial material (axis along $\hat{n}_d$, refractive indices n_o and n_e) only n_e^* depends on the incidence angle γ_d (the angle between $k_m^{(i)}$ and $\hat{n}_d$, see figure 4.9):

$$n_o^* = n_o, \qquad n_e^* = n_e^*(\gamma_d) = \frac{n_o n_e}{(n_o^2 \sin^2 \gamma_d + n_e^2 \cos^2 \gamma_d)^{1/2}}. \qquad (4.67)$$

Scattering cross-section is polarization dependent but the ordinary component ($\alpha_d = 90°$) has no dependence on the incidence angle γ_d.

For small values of $k_m R_s$ the expression of the scattering cross-section can be simplified to

$$\sigma = 2\sigma_0(k_m R_s)^2(\cos^2 \alpha_d\, s_e^2 + \sin^2 \alpha_d\, s_o^2). \qquad (4.68)$$

Exercise 2. Compute the matrix **P** for the two-dimensional phase object corresponding to a droplet with radial nematic director configuration and compare it with the same quantity for an isotropic droplet.

It is interesting to note that it has been shown [119] that in ADA the structure factor can be neglected, but has been recently taken into account by some authors [120].

4.4.2.2 Rayleigh–Gans approximation

It is assumed that each small-volume element within the droplet acts independently from others and that generates Rayleigh scattering. The simple radial and uniform droplet configurations have been studied.

For a uniform droplet the scattering cross section can be expressed as

$$\sigma = \tfrac{8}{27}\sigma_0(k_m R_s)^4[f_1(\xi,\eta) + f_2(\xi,\eta,\alpha_d,\gamma_d) + f_3(\xi,\eta,\alpha_d,\gamma_d)] \tag{4.69}$$

where

$$f_1 = (P_0 + \tfrac{1}{2}P_2)(\xi - \eta) \tag{4.70}$$

$$f_2 = (3\eta\cos^2\alpha_d\sin^2\gamma_d)[(2\xi - \eta)P_0 + (\xi - 4\eta)P_2] \tag{4.71}$$

$$f_3 = (\tfrac{27}{2}\eta^2 P_2\sin^4\gamma_d\cos^2\alpha_d) \tag{4.72}$$

$$\xi = \frac{2\varepsilon_\perp + \varepsilon_\|}{3\varepsilon_m} - 1 \tag{4.73}$$

$$\eta = \frac{\varepsilon_\| - \varepsilon_\perp}{3\varepsilon_m} \tag{4.74}$$

$$P_0 = \int_0^1 [\mu(x)]^2\, d\delta \tag{4.75}$$

$$P_2 = \int_0^1 [\mu(x)(3\cos^2\delta - 1)/2]^2\, d\delta \tag{4.76}$$

with

$$\mu(x) = 3(\sin x - x\cos x)/x^2. \tag{4.77}$$

Moreover a structure factor, valid for low droplet concentrations ($C_d < \tfrac{1}{8}$), has been derived

$$F_s = 1 - 8C_d u(2k_m R_s) \tag{4.78}$$

where C_d is droplet concentration and

$$u(x) = 3(\sin x - x\cos x)/x^3. \tag{4.79}$$

4.4.2.3 Multiple-order parameters approach

In previous sections we have defined the molecular order parameter S, equation (4.10), and the droplet order parameter S_d, equation (4.46). The first takes into account the distribution of molecular director $\hat{l}$ around the nematic director $\hat{n}$ and can be used to describe how local molecular alignment influences electric and optical LC properties: for example, refractive indices n_o and n_e are functions of S. The second takes into account the nematic director distribution inside a droplet. It is therefore reasonable to try to describe the optical properties of the whole droplet by means of S_d.

Let us define [70] the ordinary (n_{do}) and the extraordinary (n_{de}) 'droplet refractive indices' as the refractive index experienced by a wave travelling in the direction $\hat{n}_d$ or in a direction orthogonal to it. It has been shown [70, 121] that, assuming all LC molecules in the droplet are equally tilted with respect to droplet director, the droplet refractive indices are given by

$$n_{do} = \frac{2}{\pi} n_o F\left(\frac{\pi}{2}, \frac{1}{n_c}\left[\frac{2}{3}(n_e^2 - n_o^2)(1 - S_d)\right]^{1/2}\right), \tag{4.80}$$

$$n_{de} = n_e n_o [\frac{2}{3}(n_o^2 - n_e^2) + \frac{1}{3}(n_o^2 + 2n_e^2)]^{-1/2}, \tag{4.81}$$

where $F(\beta, m)$ is the first kind of complete elliptic integral.

Exercise 3. Compute the droplet refractive indices n_{de} and n_{do} for the basic droplet configurations.

For uniform droplet configuration ($S_d = 1$) we have $n_{de} = n_e$ and $n_{do} = n_o$, as expected, but for randomly oriented LC molecules ($S_d = 0$) we obtain slightly different values for n_{de} and n_{do}. However, for partially aligned droplet ($S_d > 0.5$) the error is less than 1% [70].

This allows us to compute the droplet scattering cross-section of a droplet with droplet order parameter S_d just replacing n_o and n_e with n_{do} and n_{de} in the scattering cross-section of a uniform droplet. For example if ADA can be applied, we just replace equation (4.67) with

$$n_{do}^* = n_{do}, \qquad n_{de}^* = n_{de}^*(\gamma_d) = \frac{n_{do} n_{de}}{(n_{do}^2 \sin^2 \gamma_d + n_{de}^2 \cos^2 \gamma_d)^{1/2}}. \tag{4.82}$$

Exercise 4. Compute the droplet refractive indices n_{de} and n_{do} for the basic droplet configurations.

The zero-field (no electric or magnetic field) value of the droplet order parameter (S_d^*) is mainly determined by the droplet shape, the elastic constants and the surface anchoring, and can be determined at least in simple droplet configurations. The application of an electric field changes the droplet order parameter in a way that is affected by several parameters including the zero-field droplet director and the zero-field droplet order parameter. Clearly $S_d(E=0) = S_d^*$ and it is reasonable to assume $S_d(E \to \infty) \simeq 1$, but it is almost impossible to determine the values of S_d for intermediate values of the electric field, so that only semi-empirical expressions can be used [122, 123].

4.5 Applications of PDLCs

PDLCs have a wide variety of applications due to their peculiar electro-optical and mechanical properties. Such properties allow use of PDLC in situations where other LC devices cannot be used. Here we summarize some PDLC characteristics useful in one or more applications, before describing a selected number of them in more detail in following subsections:

- PDLCs do not require rigid boundaries (glass plates) so they can be easily produced in large, flexible films.
- The amount of LCs in a PDLC film is lower than in other LC-based devices, with economic advantages since LC is an expensive material.
- Changes in the light transmission ratio are obtained by changing the light scattering cross-section within the PDLC film. In other LC devices (e.g. twisted nematic LC devices) the LC only affects light polarization state, and light transmission changes are obtained by means of dichroic polarizing sheets. The absence of polarizing sheets in PDLC devices has two related advantages: the whole light from the source is used by the device and only a moderate heating of the device occurs even under high light flux.
- PDLC films switch to a transparent state when heated over the N–I transition temperature since the isotropic LC refractive index value $n^{(I)} = n_e/3 + 2n_o/3$ is similar to the ordinary refractive index of nematic LC n_o, which is matched to polymer refractive index n_p. Such 'thermal switching effect' is often undesired, but can be used in some applications.

4.5.1 Smart windows

This is one of the first and, perhaps, the most popular PDLC application [17, 124, 125] (figure 4.4). Placing a PDLC film between two glass panes with conducting surface treatment, it is possible to switch the window between a transparent and an opalescent state by applying a low-frequency voltage across conducting electrodes, as discussed in section 4.2 (figure 4.2). Such a device can be used for privacy or light protection.

Typical values of light transmitted ratio versus applied voltage are shown in figure 4.3. Since LC droplets are anisotropic the scattering effect is both polarization-dependent and angle-dependent: typical values of light

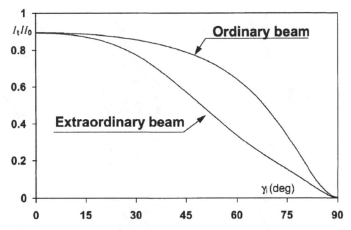

Figure 4.10. Light transmittance of a PDLC film versus incidence angle for different light polarizations. For an ordinary beam the angular dependence is mainly due to Fresnel transmission coefficient at air–glass and glass–air interfaces.

transmitted ratio versus incidence angle for ordinary and extraordinary polarized light are shown in figure 4.10.

The device we have just described, obtained using a 'positive LC' (i.e. an LC having positive dielectric anisotropy: $\Delta\varepsilon = \varepsilon_\| - \varepsilon_\perp > 0$), is transparent only when a voltage is applied (i.e. in the ON-state), but in several situations (e.g. car applications) windows are required to be transparent in the OFF-state. For this reason several types of 'reverse mode' PDLCs have been developed:

- *Negative anisotropy PDLCs.* Using a polymer-inducing homeotropic anchoring and a negative LC [78]. Producing flat droplets results in a zero-field droplet director orthogonal to the film surface. If the polymer refractive index value is close to the ordinary droplet refractive index one ($n_p = n_{do}$) the OFF-state is transparent. Since the applied electric field aligns negative LC molecules parallel to the PDLC film surface, the ON-state is opalescent.
- *Dual-frequency addressable PDLCs.* Using an LC whose dielectric anisotropy is positive at low frequency and negative at high frequency [79]. Applying a low-frequency field during polymerization, LC droplets are aligned to give a transparent film in the OFF-state. Application of a high-frequency electric field aligns LC molecules in a direction orthogonal to the applied field so that film becomes opalescent in the ON-state.
- *Homogeneously aligned, anisotropic matrix PDLCs.* Using a UV-curable LC as a polymeric matrix together with a negative LC [81]. An LC mono-acrylate, mixed with the negative LC, is homeotropically aligned (i.e. with the molecular director orthogonal to the boundary surface) between two

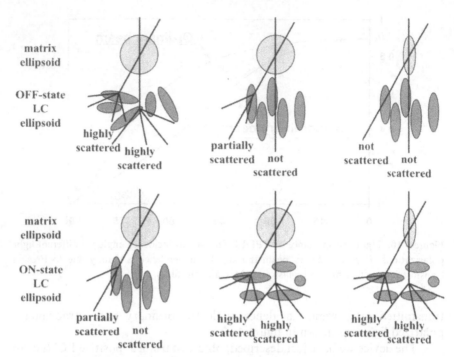

Figure 4.11. Schematic representation of the refractive index ellipsoids of polymeric matrix and LC droplets in a PDLC film, to explain angular dependence of light transmittance. Top: OFF-state; bottom: ON-state. Left: usual PDLC film (isotropic matrix, randomly oriented droplets, positive LC); middle: negative anisotropy or dual frequency addressable PDLCs (isotropic matrix, pre-aligned droplets, negative LC); right: homogeneously aligned, aniso-tropic matrix PDLC (anisotropic matrix, pre-aligned droplets, negative LC).

conducting glass plates and is photopolymerized by UV light irradiation. This preparation technique gives a preferred direction to LC molecules so that the droplet director results, orthogonal to the PDLC film surface (transparent OFF-state). Application of an electric field switches the device to an opalescent ON-state. A collateral advantage of this configura-tion is the reduced angular dependence of transmitted light intensity, since both polymeric matrix and LC droplets are anisotropic (see figure 4.11 for a schematic representation of the refractive index ellipsoids in the OFF- and ON-state).

4.5.2   Heat-resistant PDLC light modulator

High intensity (e.g. metal halide or xenon short arc) lamps commonly used on stage and in television studios have a poor light controllability because the discharge current must be kept almost constant. Light flux cannot be reduced by means of absorbing filters due to possible damage produced by light absorption and consequent temperature increase. The same is true for LC-based light modulators since they require the use of dichroic polarizers. On the contrary, temperature increase in PDLC light modulators is relatively low since non-transmitted light is scattered and not absorbed. Special care must be used in the choice of components to avoid thermal switching at temperatures above 100 °C. Both SIPS and P-PIPS have been successfully used to produce heat-resistant PDLC films to be used in high-power luminaire systems [126, 127].

4.5.3 Spatial light modulators

A spatial light modulator (SLM) is essentially an optical mask with its surface divided into zones (pixels) having independently controllable light transmittance. There are optically addressable (OA-SLM) and electronically addressable (EA-SLM) SLMs depending on how the light modulation is modified. PDLC can be used as both EA-SLM or OA-SLM and PDLC SLMs are the basis of several other PDLC applications, such as scattering mode displays or projection displays.

To obtain an OA-SLM [128] (figure 4.12) a photoconductive crystal is placed between the PDLC film and one of the transparent electrodes. The photoconductive crystal (typically BSO, i.e. $Bi_{12}SiO_{20}$) has a high electrical impedance unless it is illuminated. Therefore the electric field inside PDLC film is able to reorient LC molecules only where the substrate is illuminated. An absorbing layer and a dielectric mirror placed between the photoconductive crystal and the PDLC film allow separation of 'reading' from 'writing' beams. Such a device is able to convert a low brightness image (writing beam) into a bright image generated by a high power light source (reading beam). For this reason the term PDLC 'light valve' (PDLC-LV) is often used as a synonym.

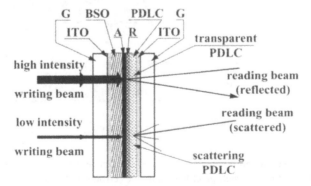

Figure 4.12. Schematic representation of an OA-SLM: G: glass; ITO: transparent electrode; BSO: photoconductive layer; A: absorbing layer (for writing beam); R: dielectric mirror (for reading beam); PDLC: polymer dispersed liquid crystal film.

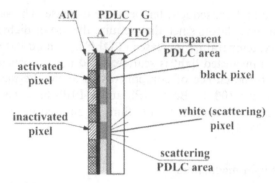

Figure 4.13. Schematic representation of an absorption-type PDLC EA-SLM with black substrate: G: glass; ITO: transparent electrode; PDLC: polymer dispersed liquid crystal film; A: absorbing layer; AM: active matrix. The absorbing layer can be removed (transmission type PDLC EA-SLM) or replaced by a dielectric mirror (reflection-type PDLC EA-SLM).

An EA-SLM (figure 4.13) can be obtained by placing the PDLC film between a passive or active matrix substrate (figure 4.14) and a transparent electrode (ITO coated glass).

A passive matrix is just a glass plate with the required number of separate electrodes, one for each pixel, each with its own conductive trace.

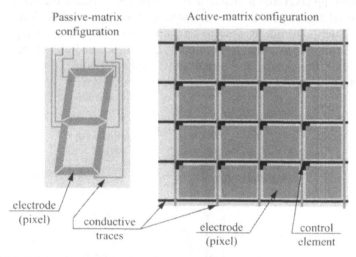

Figure 4.14. Schematic representation of passive- and active-matrix substrates. In both cases (black and white devices) each pixel has its own electrode (ITO coated area). In a colour device, not represented in this figure, each pixel corresponds to three electrodes, one for each basic colour (red, green, blue). In a passive matrix (left) each electrode has its own conductive trace. In an active matrix (right) each electrode has its own control element.

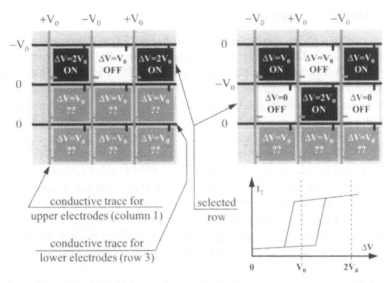

Figure 4.15. Schematic representation of a multiplexed passive matrix configuration. Inset: light transmission versus applied voltage in a large hysteresis loop PDLC film: $\Delta V = 2V_0$ or $\Delta V = 0$ are required to switch a pixel ON or OFF respectively, while $\Delta V = V_0$ leaves its state unchanged. Left: first row is selected and pixels in this row are switched ON or OFF as required. Right: second row is selected and pixels in this row are switched ON or OFF, but first row pixel status is unaffected.

Passive matrix configurations are of practical use only for small number of pixels. An active matrix is a collection of semiconductor devices, often thin film transistors (TFT) or metal insulator metal (MIM) diodes, used to selectively address each pixel. The active matrix configuration reduces the number of conductive traces at the cost of the introduction of some electronics on the SLM substrate. It is also possible to include a capacitor for each pixel to reduce the effect of voltage decay during non-addressed time.

An intermediate configuration is a multiplexed passive matrix (figure 4.15), using a PDLC with a large hysteresis loop: $\Delta V = 2V_0$ and $\Delta V = 0$ are used to switch ON and OFF respectively a pixel, while $\Delta V = V_0$ leaves unchanged the ON or OFF state of the pixel.

In both passive and active matrix configurations, selectively applying an electric field to different zones of the PDLC film allows us to selectively control light transmittance. An absorbing substrate (figure 4.13) is used to obtain dark activated pixels (absorption-type PDLC EA-SLM). Replacing the absorbing substrate with a reflective one or removing it allows us to obtain reflection- or transmission-type PDLC EA-SLMs respectively. A PDLC EA-SLM may also be combined with a thin-film electro-luminescent (TFEL) device to obtain a high visibility even in dark ambient [129].

4.5.4 Direct-view displays

Several configurations can be used to obtain direct-view displays based on an EA-SLM PDLC. One of the advantages of using PDLCs with respect to other LC configurations is obtaining flexible [130] (figure 4.16) or curved devices (no surface treatment is required and the film thickness uniformity is not a critical parameter). Another advantage is that no polarizing sheet is required, thus increasing brightness. Discussion on reflective displays using PDLC or HPDLC (section 4.5.8) and other LC/polymer composites (PSLC, PSCT) can be found in chapter 7 of a recent book [131].

The simplest device is obtained using a PDLC EA-SLM in absorption-type configuration (figure 4.13): OFF-state pixels are white due to scattered light, ON-state pixels are black due to absorption of black substrate. Contrast ratio is generally low since only backscattering is used. An air gap between PDLC film and light-absorbing substrate (not illustrated in the figure) increases overall backscattering (due to reflection at PDLC air interface).

Better results can be obtained adding a dichroic dye within PDLC droplets and placing the EA-SLM over a coloured reflective surface (figure 4.17). OFF-state pixels are black due to light absorption within droplets. ON-state pixels are coloured since PDLC is transparent and the background selectively reflects light.

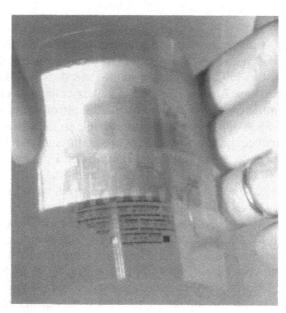

Figure 4.16. A PDLC based display on flexible polymeric substrate. From C D Sheraw *et al* 2002 *Appl. Phys. Lett.* **80** 1088.

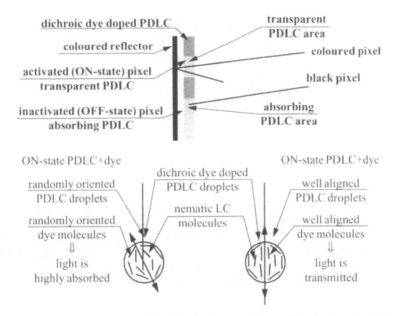

Figure 4.17. Schematic representation of a display based on a dichroic dye-doped PDLC. Dye and LC molecules are aligned to each other. In ON-state dye molecules are randomly oriented so that light is partially scattered and highly absorbed. As a consequence inactivated pixels are black. In the OFF-state LC and dye molecules are aligned to the applied field so that light is not scattered and slightly absorbed. As a consequence light reaches the coloured reflector and activated pixels are coloured.

4.5.5 Projection displays

Projection displays are used to reproduce images on large screens. PDLC-based devices are compact, low-weight and, since PDLC do not require use of polarizers, produce images bright enough to allow use in normally lighted rooms.

In a transmission-type [132] PDLC projection display (figure 4.18) a white light source (W) is collimated over a transmission type PDLC EA-SLM (P) and then projected onto the screen (S). Only light passing through ON-pixels (transparent) is collected by a lens (L) and focused by the objective (O) on the screen (S) thus producing a bright spot. Light impinging on OFF-pixels is scattered and is blocked by an absorbing aperture (A). Colour images are obtained using three PDLC SLMs for red, green and blue light respectively. This has a collateral advantage, with respect to direct-view displays: the number of pixels in each SLM is reduced by one-third. The three basic colours are separated by means of a set of dichroic mirrors (D), modulated by the SLMs (P) and then recombined by means of another set of dichroic mirrors (D).

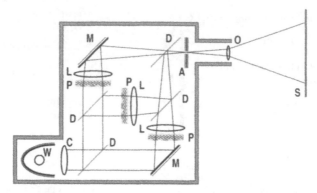

Figure 4.18. Schematic representation of a transmission-type PDLC projection display. W: white light source; C: collimator lens system; D: dichroic mirrors used both to separate and to recombine red, green and blue component of light; M: mirrors; P: transmission type PDLC EA-SLMs (one for each basic colour); L: lenses used to collect transmitted light; A: absorbing diaphragm used to reject scattered light; O: objective; S: screen. The electronic circuitry required to drive the EA-SLMs is not illustrated.

With a different configuration [133, 134] (figure 4.19), and making use of a dichroic prism (D_P), it is possible to use reflection type PDLC-SLMs (figure 4.13) with a higher total light efficiency [134].

It is also possible to use a PDLC-LV (i.e. a PDLC OA-SLM) [135] (figure 4.20): a low-intensity light source (W_L) and three small-area LCDs are used to create three images on the rear surface of three PDLC-LVs (P). All three images are grey-level, but each one corresponds to a different basic colour (red, green and blue). A high-intensity light source (W_H), decomposed into red, green and blue components, is used to illuminate the front surface on

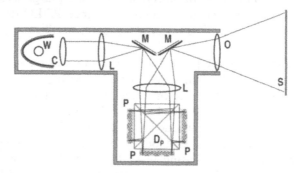

Figure 4.19. Schematic representation of a reflection type PDLC projection display. W: white light source; C: collimator lens system; D_P: dichroic prism used both to separate and to recombine red, green and blue component of light; M: mirrors; P: reflection type PDLC EA-SLMs (one for each basic colour); L: lenses; O: objective; S: screen. The electronic circuitry required to drive the EA-SLMs is not illustrated.

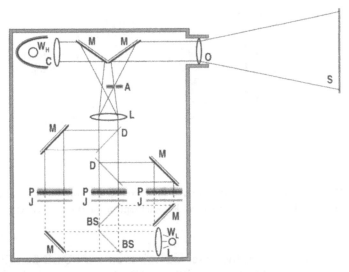

Figure 4.20. Schematic representation of a 'light valve'-type PDLC projection display. W_L: low intensity white light source; W_H: high-intensity white light source; C: collimator lens system; BS: beam splitters; J: device used to generate separate grey-level images corresponding to red, green and blue components; D: dichroic mirrors used to recombine red, green and blue component of light; M: mirrors; P: PDLC-LVs (one for each basic colour); L: lenses; A: absorbing diaphragm used to reject scattered light; O: objective; S: screen. The electronic circuitry required to generate separate red, green and blue images and to drive the LCDs is not illustrated.

the PDLC-LVs. As described above in section 4.5.3, the front surface of the PDLC-LV acts as a scattering surface in dark areas and as a reflecting surface in illuminated areas. Only reflected light is collected by a lens (L) and focused by the objective (O) onto the screen (S), while scattered light is absorbed by a diaphragm (A). With this technique a contrast ratio of 80:1 and a brightness greater than 1800 ANSI lumen has been obtained [135].

Recently [136] a configuration with 'lasing pixels' has been proposed. The PDLC is used as a variable loss element within an optically pumped high-gain laser cavity (figure 4.21). Areas of the laser cavity corresponding to OFF-state PDLC pixels are below lasing threshold so that no lasing occurs. This technique allows high output intensity to be obtained ($600\,\mathrm{lumen/cm^2}$) since OFF-state pixels are not required to dissipate energy, and high contrast (1000:1) since the light emitted by non-lasing pixels is practically zero.

4.5.6 Eye-protection viewer

An interesting application of the 'thermal switching effect' of a PDLC film is an eye-protection viewer device (figure 4.22) [83]. An optical system (L_1)

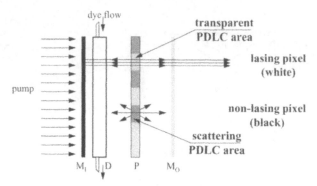

Figure 4.21. Schematic representation of a 'lasing pixel' device. M_I: input dichroic mirror; D: dye cell; P: PDLC acting as a variable loss element in the optical cavity; M_O: output coupling mirror.

generates a real image of the observed object onto a PDLC film (P) placed over a black-coated (A) heat dissipator (H). A second optical system (M and L_2) converts this real image into a virtual image viewed by the observer's eye (E). If the observer looks towards an intense light source (e.g. a laser beam) the PDLC portion where the image of the source produced becomes transparent due to N–I LC transition. As a consequence, intense and potentially eye-dangerous light is absorbed by the black background and the transported energy is dissipated. The observer sees the image with a black spot in place of the intense light source. This protective devices has the advantage of being selective (only the 'dangerous' part of the image is 'removed') and reversible (LC returns to nematic mesophase when the intense light source is turned off). The main disadvantage is that only a fraction of the light impinging on the PDLC is used to generate the observable virtual image.

4.5.7 Sensors

Several sensor types can be obtained with special PDLC configurations. A 'partially exposed' configuration (i.e. a SIPS PDLC film deposited, or

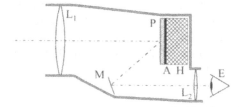

Figure 4.22. Schematic representation of an eye-protection device based on the 'thermal switching effect' of a PDLC film: L_1, L_2: lenses; M: mirror; P: polymer dispersed liquid crystal film; A: light absorbing surface; H: heat dissipator; E: observer's eye.

sprayed, on a glass plate and not covered with a second glass plate) has been proposed as a gas-flow sensor [24]. Droplets on the exposed surface of the thin (2–5 μm) PDLC film respond to an air flow changing droplet director orientation (shear stress induced director reorientation) [137]. The changes in droplet configuration can be detected observing the PDLC film between crossed polarizers, thus allowing skin friction measurement [26].

Placing a PDLC film between a dielectric substrate and a conducting surface [138] it is possible to measure two-dimensional charge distributions on dielectric surfaces in real time. The electric field generated by the local charge density affects PDLC light transmittance which is detected by means of a CCD camera. A P-PIPS PDLC film has been used to detect a charge distribution in 333 ms with 30 μm spatial resolution and 5×10^{-12} C/mm^2 charge sensitivity.

An electric field sensor has been realized with a PDLC-coated optical fibre [139]. In a step-index multimode fibre with a 200 μm silica core the polymeric cladding is removed for a short length, and is replaced by a TIPS PDLC. In usual optical fibre operation an evanescent electromagnetic field is generated in the cladding. Replacing the cladding with a PDLC affects the evanescent field and therefore the light propagation along the fibre. The electric field changes the optical properties of the PDLC and, as a consequence, the transmission output of the guided light.

Temperature sensors can be realized using a dual frequency LC [51, 73]. Since such LCs behave as positive ($\Delta\varepsilon > 0$) for low frequency and as negative ($\Delta\varepsilon < 0$) for high frequencies there is some frequency ν_c for which $\Delta\varepsilon = 0$. Dielectric constants, and therefore ν_c, are a function of the temperature. If the PDLC film is operated at frequency ν_c (T_0) a small temperature variation switches PDLC film from a transparent to an opalescent state or vice versa.

4.5.8 Holographic PDLCs

A holographic-PDLC (HPDLC) is obtained (figure 4.23) if polymerization of a P-PIPS PDLC occurs in a spatially-modulated light, e.g. a grating produced by two coherent laser beams [140, 141]. Polymerization is faster in high-light intensity zones so that LC droplets are mainly confined to low-light intensity zones [142]. This technique allows us to realize an electrically switchable holographic grating. In the OFF-state (no electric field) the HPDLC film acts as reflection (or transmission) grating, due to the difference in the refractive index of polymer planes and LC reach planes. In the ON-state (a suitable electric field applied across the film) the refractive index mismatch disappears and the HPDLC film acts as transparent film. According to Bragg's law an OFF-state HPDLC film selectively reflects or transmits light of specific wavelength. A HPDLC is therefore an electrically

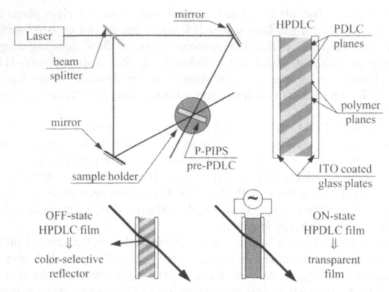

Figure 4.23. Holographic polymer dispersed liquid crystal preparation technique: a P-PIPS PDLC film is cured in the space region where interference fringes are produced by means of two laser beams. Polymerization starts in the bright fringes so that LC diffuses out into droplets only in dark fringes. As a final result, polymer planes are alternated with LC reach planes: OFF-state HPDLC film acts as a diffraction grating. Applying an electric field across the film, the refractive index mismatch disappears: ON-state HPDLC acts as a transparent film.

switchable colour selective reflector useful for several interesting applications [143].

Colour reflective displays have been obtained both stacking two memory-type, obtained using a smectic LC, HPDLC [144] or recording multiple gratings within the same HPDLC film [145]. Other applications are integrated lenses and filters for microdisplays or optical switches for fibre switching in telecommunication systems [146, 147] or photonic delay lines [148].

4.5.9 Scattering polarizers

The polarization dependence of light scattered by a PDLC film has suggested the possibility of making light polarizers which, differently from dichroic ones, scatter instead of absorbing unwanted light component. This has been achieved in two different ways: stretching PDLC film after polymerization [59, 86, 87] to obtain elongated droplets and therefore a zero-field film director parallel to the film surface; or applying a transversal [88, 89] electric field. Scattering-based polarizers have the advantage, with respect

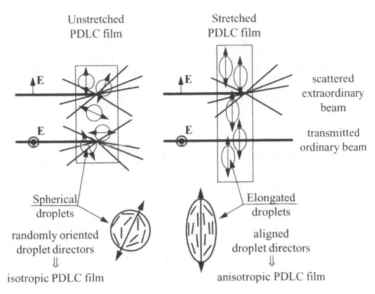

Figure 4.24. Schematic representation of a stretched PDLC polarizer. In a non-stretched OFF-state PDLC film (left) light is scattered independently of its polarization. After stretching (right) ordinary light beam (bottom) does not recognize any refractive index mismatch so that it is transmitted, while extraordinary light beam (top) is scattered.

to dichroic ones, that the unwanted light component is scattered instead of being absorbed, thus reducing energy absorption and consequent device heating.

In stretched PDLC polarizers [59, 86, 87] an emulsion-type PDLC film is unidirectionally stretched up to 300% the initial length. In unstretched film the almost spherical droplets assume the bipolar configuration with droplet directors randomly oriented. After stretching, the droplets assume the shape of prolate ellipsoids with the major axis in the stretching direction. As a consequence droplet directors are mainly oriented in the same direction (it has been found that the zero-field film order parameter S_f^* is practically 1 for 100% stretching). Therefore the stretched PDLC film acts as a polarizing film since only an ordinary beam is transmitted without scattering (figure 4.24).

A different approach [88, 89] uses a transversal electric field (figure 4.25) obtained by embedding metallic spacers within the PDLC film. In OFF-state, droplets are randomly oriented and therefore light is highly scattered, independently of its polarization direction. On the contrary, in the ON-state, droplet directors are all aligned to each other in a direction lying in the PDLC film plane. As a consequence only ordinary polarized light is highly scattered: the device behaves as an electrically controllable polarizer.

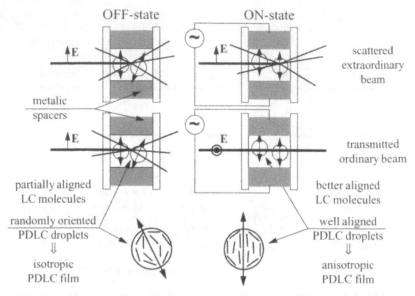

Figure 4.25 Schematic representation of a transversal field PDLC polarizer. In the OFF-state PDLC film (left) light is scattered independently of its polarization. In the ON-state, obtained applying a transversal electric field by means of metallic spacers embedded in the PDLC film (right), ordinary light beam (bottom) does not recognize any refractive index mismatch so that it is transmitted, while extraordinary light beam (top) is scattered.

References

[1] de Gennes P 1992 *Rev. Mod. Phys.* **64** 645–648
[2] Kohn K 1999 *Rev. Mod. Phys.* **71** S59–S77
[3] Witten T A 1999 *Rev. Mod. Phys.* **71** S367–S373
[4] Stephen M J and Straley J P 1974 *Rev. Mod. Phys.* **46** 617–704
[5] de Gennes P G and Prost J 1993 *The Physics of Liquid Crystals* (Oxford: Clarendon Press)
[6] de Jeu W H 1979 *Physical Properties of Liquid Crystalline Materials* (New York: Gordon and Breach)
[7] Priestley E B, Wojtowicz P J and Sheng P (eds) 1979 *Introduction to Liquid Crystals* (New York: Plenum)
[8] Singh S 2000 *Phys. Rep.* **324** 107–269
[9] Dubois-Violette E and Parodi O 1969 *J. Phys. Coll.* **30/C4** 57–64
[10] Beni G, Craighead H G and Hackwood S 1981 *Appl. Phys. Lett.* **39** 195–197
[11] Craighead H G, Cheng J and Hackwood S 1982 *Appl. Phys. Lett.* **40** 22–25
[12] Fergason J L 1984 US Patent 4 435 047
[13] Fergason J L 1985 *Digest SID* **85** 68–70
[14] Doane J W, Vaz N A, Wu B G and Zumer S 1986 *Appl. Phys. Lett.* **48** 269–271
[15] Doane J W, Chidichimo G and Vaz N A P 1987 US Patent 4 688 900

[16] NCAP is a trademark registered in 1994 in the USA by Raychem Corporation, Delaware, California, as 'liquid crystal film for producing character and/or graphic type displays'.

[17] Drzaic P S 1986 *J. Appl. Phys.* **60** 2142–2148

[18] Tomita A and Jones P J 1992 *Proc. SPIE* **1665** 274–279

[19] Reamey R H, Montoya W and Wartenberg M 1991 *Proc. SPIE* **1455** 39–44

[20] Jones P J, Tomita A and Wartenberg M 1991 *Proc. SPIE* **1456** 6–14

[21] Reamey R H, Montoya W and Wong A 1992 *Proc. SPIE* **1665** 2–7

[22] Havens J R, Ishioka J, Jones P J, Lau A, Tomita A, Asano A, Konuma N, Sato K and Takemoto I 1997 *Proc. SPIE* **3013** 185–192

[23] Ono H and KaWatsuki N 1994 *Jpn. J. Appl. Phys.* **33** L1778–L1781.

[24] Parmar D S and Sigh J J 1992 *Appl. Phys. Lett.* **61** 2039–2041

[25] Simoni F, Bloisi F and Vicari L 1992 *Mol. Cryst. Liq. Cryst.* **223** 169–179

[26] Parmar D S and Holmes H K 1993 *Rev. Sci. Instrum.* **64** 538–541

[27] Shen C and Kyu T 1995 *J. Chem. Phys.* **102** 556–562

[28] Zumer S, Crawford G P and Doane J W 1995 *Mol. Cryst. Liq. Cryst.* **261** 577–592

[29] Palffy-Muhoray P, Lee M A and West J L 1990 *Mol. Cryst. Liq. Cryst.* **179** 445–460

[30] Li L, Yuan H Y and Palffy-Muhoray P 1991 *Mol. Cryst. Liq. Cryst.* **198** 239–246

[31] Simoni F, Bloisi F and Vicari L 1993 *Int. J. Nonlin. Opt. Phys.* **2** 353–365

[32] Lu Z-J and Yang D-K 1994 *Appl. Phys. Lett.* **65** 505–507

[33] Komitov L, Rudquist P, Aloe G and Chidichimo G 1994 *Mol. Cryst. Liq. Cryst.* **251** 317–347

[34] Kitzerow H-S, Molsen H and Heppke G 1992 *Appl. Phys. Lett.* **60** 3093–3095

[35] Pozhidaev E P, Smorgon S L, Andreev A L, Kompanets I N, Zyryanov V Ya and Shin S T 1997 *OSA TOPS* **14** 94–101

[36] Murai H, Gotoh T, Nakata T and Hasegawa E 1997 *J. Appl. Phys.* **81** 1962–1965

[37] Tanaka K, Kato K and Date M 1999 *Jpn. J. Appl. Phys.* **38** L277–L278

[38] Kato K, Hisaki T and Date M 1999 *Jpn. J. Appl. Phys.* **38** 1466–1469

[39] Escuti M J, Kossyrev P, Crawford G P, Fiske T G, Colegrove J and Silverstein L D 2000 *Appl. Phys. Lett.* **77** 4262–4264

[40] Vilfan M, Zalar B, Fontecchio A K, Vilfan M, Escuti M J, Crawford G P and Zumer S 2002 *Phys. Rev. E* **66** 021710 (9 pages)

[41] Dierking I, Kosbar L L, Afzali–Ardakani A, Lowe A C and Held G A 1997 *J. Appl. Phys.* **81** 3007–3014

[42] Dierking I, Kosbar L L, Afzali–Ardakani A, Lowe A C and Held G A 1997 *Appl. Phys. Lett.* **71** 2454–2456

[43] Vicari L 1999 *J. Opt. Soc. Am. B* **16** 1135–1138

[44] Simoni F, Basile F, Bloisi F, Vicari L and Aliev F 1994 *Mol. Cryst. Liq. Cryst.* **239** 257–261

[45] Yang D K, Chien L C and Doane J W 1992 *Appl. Phys. Lett.* **60** 3102–3104

[46] Fung Y K, Yang D K and Doane J W 1992 *Proc. SPIE* **1664** 41–47

[47] Bouteiller L and Le Barny P 1996 *Liq. Cryst.* **21** 157–174

[48] Drzaic P S 1995 *Liquid Crystal Dispersions* (Singapore: World Scientific)

[49] Zhang G M, Hong Z, Changxing Z, Wu B G and Lin J W 1992 *Proc. SPIE* **1815** 233–237

[50] Vaz N A and Montgomery Jr G P 1989 *J. Appl. Phys.* **65** 5043–5050

[51] Jain S C, Thakur R S and Lakshmikumar S T 1993 *J. Appl. Phys.* **73** 3744–3748

[52] Hasegawa R, Sakamoto M and Sasaki H 1993 *Appl. Spectrosc.* **47** 1386–1389

[53] McFarland C A, Koenig J L and West J L 1993 *Appl. Spectrosc.* **47** 321–329
[54] Li Z, Kelly J R, Palffy-Muhoray P and Rosenblatt C 1992 *Appl. Phys. Lett.* **60** 3132–3134
[55] Drzaic P S 1988 *Liq. Cryst.* **3** 1543–1559
[56] West J L, Fredley D S and Carrel J C 1992 *Appl. Phys. Lett.* **61** 2004–2005
[57] Drzaic P S and Muller A 1989 *Liq. Cryst.* **5** 1467–1475
[58] Drzaic P S 1991 *Mol. Cryst. Liq. Cryst.* **198** 61–71
[59] Aphonin O A, Panina Yu V, Pravdin A B and Yakovlev D A 1993 *Liq. Cryst.* **15** 395–407
[60] Ono H and Kawatsuki N 1995 *Jpn. J. Appl. Phys.* **34** 1601–1605
[61] Kim J Y, Cho C H, Palffy-Muhoray P, Mustafa M and Kyu T 1993 *Phys. Rev. Lett.* **71** 2232–2236
[62] West J L 1988 *Mol. Cryst. Liq. Cryst. Inc. Nonlin. Opt.* **157** 427–441
[63] Fuh A and Caporaletti O 1989 *J. Appl. Phys.* **66** 5278–5284
[64] Margemum J D, Yamagishi F G, Lackner A M, Sherman E, Miller L J and van Ast C I 1993 *Liq. Cryst.* **14** 345–350
[65] Bloisi F, Ruocchio C and Vicari L 1997 *J. Phys. III (France)* **7** 1097–1102
[66] Vaz N A and Montgomery Jr J P 1987 *J. Appl. Phys.* **62** 3161–3172
[67] Wu B G, West J L and Doane JW 1987 *J. Appl. Phys.* **62** 3925–3931
[68] Erdmann J H, Zumer S and Doane J W 1990 *Phys. Rev. Lett.* **64** 1907–1910
[69] Simoni F, Cipparrone G and Umeton C 1990 *Appl. Phys. Lett.* **57** 1949–1951
[70] Basile F, Bloisi F, Vicari L and Simoni F 1993 *Phys. Rev. E* **48** 432–438
[71] Kelly J R and Wu W 1993 *Liq. Cryst.* **14** 1683–1694
[72] Nolan P, Tillin M and Coates D 1993 *Liq. Cryst.* **14** 339–344
[73] Jain S C and Thakur R S 1992 *Appl. Phys. Lett.* **61** 1641–1642
[74] Ondris-Crawford R, Boyko E P, Wagner B G, Erdman J H, Zumer S and Doane J W 1991 *J. Appl. Phys.* **69** 6380–6386
[75] Wu B G, Erdmann J H and Doane J W 1989 *Liq. Cryst.* **5** 1453–1465
[76] Jain S C and Rout D K 1991 *J. Appl. Phys.* **70** 6988–6992
[77] West J L, Kelly J R, Jewell K and Ji Y 1992 *Appl. Phys. Lett.* **60** 3238–3240
[78] Ma Y D and Wu B G 1990 *Proc. SPIE* **1257** 46–57
[79] Gotoh T and Murai H 1992 *Appl. Phys. Lett.* **60** 392–394
[80] Klosowicz S J, Nowinowski-Kruszelnicki E, Zmija J and Dabrowski R 1995 *Proc. SPIE* **2372** 363–366
[81] Yamaguchi R, Waki Y and Sato S 1997 *Jpn. J. Appl. Phys.* **36** 2771–2774
[82] McLaughlin C W, Drzaic P and Marsland S 1986 US Patent 4 749 261
[83] Land P L and Schmitt M G 1995 US Patent 5 448 382
[84] Vicari L 1997 *J. Appl. Phys.* **81** 6612–6615
[85] Kato K, Hisaki T and Date M 1999 *Jpn. J. Appl. Phys.* **38** 805–808
[86] Aphonin O A 1995 *Liq. Cryst.* **19** 469–480
[87] Aphonin O A 1996 *Mol. Cryst. Liq. Cryst.* **281** 105–122
[88] Bloisi F, Ruocchio C, Terrecuso P and Vicari L 1996 *Liq. Cryst.* **20** 377–379
[89] Bloisi F, Terrecuso L and Vicari L 1997 *J. Opt. Soc. Am. A* **14** 662–668
[90] Oton M J, Serrao A, Serna C J and Levy D 1991 *Liq. Cryst.* **10** 733–739
[91] Chang W P, Whang W T and Wong J C 1995 *Jpn. J. Appl. Phys.* **34** 1888–1894
[92] Vissenberg M C J M, Stallinga S and Vertogen G 1997 *Phys. Rev. E* **55** 4367–4377
[93] Govers G and Vertogen G 1984 *Phys. Rev. A* **30** 1998–2000
[94] Stallinga S and Vertogen G 1994 *Phys. Rev. E* **49** 1483–1494

[95] Magnuson M L, Fung B M and Bayle J P 1995 *Liq. Cryst.* **19** 823–832

[96] Nehring J and Saupe A 1971 *J. Chem. Phys.* **54** 337–343

[97] Ondris-Crawford R J, Crawford G P, Zumer S and Doane J W 1993 *Phys. Rev. Lett.* **70** 194–197

[98] van de Hulst H C 1957 *Light Scattering by Small Particles* (New York: Wiley)

[99] Bohren C F and Huffman D R 1983 *Absorption and Scattering of Light by Small Particles* (New York: Wiley)

[100] Zumer S 1988 *Phys. Rev. A* **37** 4006–4015

[101] Van de Hulst 1980 *Multiple Light Scattering: Tables, Formulas and Applications* vol I and II (New York: Academic)

[102] Zumer S, Golemme S and Doane J W 1989 *J. Opt. Soc. Am. A* **6** 403–411

[103] Zumer S and Doane J W 1986 *Phys. Rev. A* **34** 3373–3386

[104] Williams R D 1986 *J. Phys. A: Math. Gen.* **19** 3211–3222

[105] Lebwohl P A and Lasher G 1972 *Phys. Rev. A* **6** 426–429

[106] Lasher G 1972 *Phys. Rev. A* **5** 1350–1354

[107] Zannoni C 1986 *J. Chem. Phys.* **84** 424–433

[108] Chiccoli C, Pasini P, Semeria F and Zannoni C 1990 *Phys. Lett. A* **150** 311–314

[109] Berggren E, Zannoni C, Chiccoli C, Pasini P and Semeria F 1994 *Phys. Rev. E* **49** 614–622

[110] Chiccoli C, Pasini P, Semeria F and Zannoni C 1992 *Mol. Cryst. Liq. Cryst.* **221** 19–28

[111] Berggren E, Zannoni C, Chiccoli C, Pasini P and Semeria F 1994 *Phys. Rev. E* **50** 2929–2939

[112] Chiccoli C, Pasini P, Skacej G, Zannoni C and Zumer S 2000 *Phys. Rev. E* **62** 3766–3774

[113] Margerum J D, Lackner A M, Ramos E, Lim K-C and Smith Jr W H 1989 *Liq. Cryst.* **5** 1477–1487

[114] Golemme A, Zumer S, Doane J W and Neubert M E 1998 *Phys. Rev. A* **37** 559–569

[115] Kelly J and Palffy-Muhoray P 1994 *Mol. Cryst. Liq. Cryst.* **243** 11–29

[116] Basile F, Bloisi F, Vicari L and Simoni F 1994 *Mol. Cryst. Liq. Cryst.* **251** 271–281

[117] Rout D K and Jain S C 1992 *Mol. Cryst. Liq. Cryst.* **210** 75–81

[118] Lin H, Ding H and Kelly J R 1995 *Mol. Cryst. Liq. Cryst.* **262** 99–109

[119] Whitehead Jr J B, Zumer S and Doane J W 1993 *J. Appl. Phys.* **73** 1057–1065

[120] Dick V P and Loiko V A 2001 *Liq. Cryst.* **28** 1993–1998

[121] Bloisi F, Terrecuso P, Vicari L and Simoni F 1995 *Mol. Cryst. Liq. Cryst.* **266** 229–239

[122] Bloisi F and Vicari L 1999 *Recent Res. Dev. Macromol. Res.* **4** 113–127

[123] Bloisi F and Vicari L 2003 *Optics and Lasers in Engineering* **39** 389–408

[124] Lampert C M 1998 *Solar Energy Materials and Solar Cells* **52** 207–221

[125] Sanchez Pena J M, Vazquez C, Perez I, Rodriguez I and Oton J M 2002 *Opt. Eng.* **41** 1608–1611

[126] Fujikake H, Tanaka Y, Kimura S, Asakawa H, Tamura T, Kita H, Takeuchi K, Ogawa H, Nagashima A, Utsumi Y and Takizawa K 2000 *Jpn. J. Appl. Phys.* **39** 5870–5874

[127] Takizawa K, Fujii T, Fujikake H, Hirabayashi T, Tanaka Y, Hara K, Takano S, Asakawa H and Kita H 1999 *Appl. Opt.* **38** 2570–2578

[128] Takizawa K, Kikuchi H, Fujikake H, Namikawa Y and Tada K 1994 *Jpn. J. Appl. Phys.* **33** 1346–1351

[129] Fuh A Y-G, Huang C-Y, Shen C-R, Lin G-L and Tsai M-S 1994 *Jpn. J. Appl. Phys.* **33** L870–L872

[130] Sheraw C D, Zhou L, Huang J R, Gundlach D J, Jackson T N, Kane M G, Hill I G, Hammond M S, Campi J, Greening B K, Francl J and West J 2002 *Appl. Phys. Lett.* **80** 1088–1090

[131] Wu S-T and Yang D-K 2001 *Reflective Liquid Crystal Displays* (New York: Wiley)

[132] Shikama S, Kida H, Daijogo A, Okamori S, Ishitani H, Maemura Y, Kondo M, Murai H and Yuki M 1995 *Digest SID* **95** 231–234

[133] Glueck J, Ginter E, Lueder E and Kallfass T 1995 *Digest SID* **95** 235–238

[134] Nagae Y, Ando K, Asano A, Takemoto I, Havens J, Jones P, Reddy D and Tomita A 1995 *Digest SID* 95 223–226

[135] Kikuchi H, Fujii T, Kawakita M, Fujikake H and Taktzawa K 2000 *Opt. Eng.* **39** 656–669

[136] Firehammer J A, Crawford G P aand Lawandy N M 1998 *Appl. Phys. Lett.* **73** 590–592

[137] Parmar D S and Singh J J 1993 *Liq. Cryst.* **14** 361–369

[138] Budakian R and Putterman S J 2000 *Rev. Sci. Instrum.* **71** 444–449

[139] Tabib-Azar M, Sutapun B, Srikhirin T, Lando J and Adamovsky G 2000 *Sensors and Actuators* **84** 134–139

[140] Sutherland R L 1991 *J. Opt. Soc. Am. B* **8** 1516–1525

[141] Shuterland R L, Tondiglia V P, Natarajan L V, Bunning T J and Adams W W 1994 *Appl. Phys. Lett.* **64** 1074–1076

[142] Bunning T J, Natarajan L, Tondiglia V and Sutherland R L 1995 *Polymer* **36** 2699–2708

[143] Sutherland R L, Natarajan L V, Tondiglia V P and Bunning T J 1998 *Proc. SPIE* **3421** 8–18

[144] Date M, Takeuchi Y and Kato K 1998 *J. Phys. D: Appl. Phys.* **31** 2225–2230

[145] Fontecchio A K, Bowley C C, Chmura S M, Le Li, Faris S and Crawford G P 2001 *J. Opt. Technol.* **68** 652–656

[146] Smith R T, Popovich M M and Sagan S F 2000 *Proc. SPIE* **4207** 31–38

[147] Jihong Z, Lingjuang G and Songlin Z 2002 To be published.

[148] Madamopoulos N and Riza N A *Opt. Commun.* 225–237

Chapter 5

New developments in photo-aligning and photo-patterning technologies: physics and applications

V G Chigrinov, V M Kozenkov and H S Kwok

5.1 Introduction

Photo-alignment possesses obvious advantages in comparison with the usual 'rubbing' treatment of the substrates of liquid crystal display (LCD) cells. Possible benefits for using this technique include:

(i) the elimination of electrostatic charges and impurities as well as mechanical damage of the surface;
(ii) the possibility of producing structures with the required LC director azimuth within the selected areas of the cell, thus allowing pixel division to improve viewing angles;
(iii) the potential increase of manufacturing yield, especially in LCDs with active matrix addressing, where fine tiny pixels of a high resolution LCD screen are driven by thin film transistors on a silicone substrate.

The effect of LC photoaligning is a direct consequence of the appearance of the photo-induced optical anisotropy and dichroism absorption in thin amorphous films, formed by molecular units with anisotropic absorption properties [1]. The first publication on LC photo-alignment appeared in 1988 and discussed the application of the reversible *cis–trans* isomerization of the azo-benzene molecular layers attached to the solid surface to the switching of the alignment of the adjacent LC layer from homeotropic to azimuthally random planar orientation [2]. The optical control of LC alignment was made by changing the wavelength of the non-polarized light illumination [2]. Later it was shown that the alignment of a liquid crystal medium could be made by illuminating a dye-doped polymer alignment layer with a polarized light [3]. LC molecules in contact with the illuminated area were homogeneously aligned perpendicular to the direction of the laser polarization and remain aligned in the absence of the laser light. Very soon the LC photo-alignment procedure was made using cinnamoyl side-chain polymers [4, 5] and polyimide aligning agents [6]. The area of LC photo-alignment is very rapidly developing and the vast majority of the new materials, techniques and LCD prototypes based on photo-alignment technology appeared recently, which was partially covered by some recent review articles [7–10].

In this review we shall discuss:

  (i) the fundamental mechanisms of LC photo-alignment with proper examples of the materials for their realization;
 (ii) LC–surface interaction and methods to achieve a high quality LC alignment with a controllable anchoring energy of LC onto the photo-aligning layer;
(iii) pretilt angle generation in photo-aligned LC cells;
 (iv) multi-domain LC cells and photo-patterning technology for phase retarders, birefringent colour filters and superthin internal polarizers, security applications;
  (v) photo-aligning of ferroelectric LC, vertical aligned nematic (VAN) modes and discotic LC.

We shall also provide some examples of novel LC devices, based on photo-aligning and photo-patterning technology.

5.2   Mechanisms of LC photo-alignment

The progress in application of LC photo-aligning for LCDs has stimulated many fundamental studies of its mechanism. However, an adequate explanation of the photo-alignment phenomena is still absent. Several publications appeared recently and provide only a qualitative explanation of the phenomenon of photo-alignment. A more adequate explanation of the process is still in progress, taking into account both photo-chemical and photo-physical transformations in photosensitive layers under the action of a polarized or non-polarized, but directed light.

We can distinguish the following mechanisms of the photo-alignment:

  (i) photochemical reversible *cis–trans* isomerization in azo-dye containing polymers, monolayers and pure dye films;
 (ii) pure photophysical reorientation of the azo-dye chromophore molecules or azo-dye molecular solvates due to the diffusion under the action of polarized light;
(iii) topochemical crosslinking in cinnamoyl side-chain polymers;
 (iv) photodegradation in polyimide materials.

The first two processes present the reversible transformations, while the latter two can be referred to as irreversible photo-chemical phenomena, taking place in photosensitive aligning films. For *cis–trans* isomerization the change of the absorption spectra is observed after illumination [11], which is not the case for the photo-stable azo-dye molecules, involved in a reorientation and solvate formation process under the action of polarized light [12]. Let us consider the mechanisms in more detail and illustrate them on various photo-aligning materials.

5.2.1 *Cis–trans* isomerization

5.2.1.1 *Command surface*

Under the action of a polarized light a reversible *trans–cis* transformation of azo-dye molecules can be observed [2]. Ichimura *et al* showed that if the dye molecules are directly attached to the surface, then a so-called 'command surface' can be created [2, 13–15] (figure 5.1). The conformation of the azobenzene dye units was shown to undergo reversible *cis–trans* transformations under the subsequent UV–visible non-polarized light illumination. Thus the ensemble of the liquid crystal (LC) molecules will reversibly change its orientation from the homeotropic to random planar and vice versa, so that one surface dye unit may cause a reorientation of about one

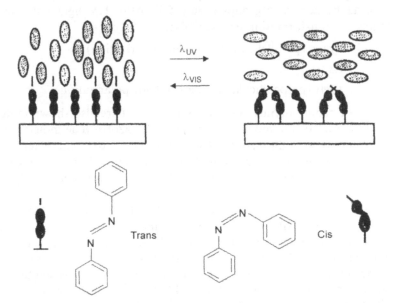

Figure 5.1. Reversible homogeneous to homeotropic transition in liquid crystal cell, caused by *cis–trans* isomerization of the azobenzene units, attached at the glass substrate ('command surface'). UV light ($\lambda = 365\,\text{nm}$) transforms the azobenzene units to the *cis* form (homogeneous LC alignment), while visible light ($\lambda = 440\,\text{nm}$) restores the *trans* form configuration [2].

million LC molecules in the bulk. Such a surface with a reversible *cis–trans* isomerization taking place under the action of UV–visible light transformation was called 'command surface' [15]. However, this method of controlling the LC reorientation reveals certain disadvantages, such as:

(i) homeotropic to planar configuration is azimuthally degenerated;
(ii) the dark relaxation of the *cis* configuration is observed, thus spontaneously restoring the initial homeotropic LC configuration;
(iii) the number of reversible cycles is limited by a photochemical stability of the dye layers.

Ichimura *et al* published a number of papers, showing that the effect of *cis–trans* isomerization is very important in most cases [2, 13–15]. The application of photo-optical regulation of LC orientation in waveguide structures was also proposed [16]. The new silane [17] and stilbene [18] surfactants were used to produce the photo-regulated alignment.

5.2.1.2   Cis–trans transformations in azo-dye side chain polymers and azo-dye in polymer matrix

The concept of the *cis–trans* isomerization in azo-dye side chain polymers was developed by Shibaev *et al* [19]. The side chain of the polymer was made of azobenzene side groups (figure 5.2). After UV–light illumination the following transformation occurs [20] (figure 5.2):

> Trans (parallel to the UV light polarization) ⇒
>
> *cis* ⇒
>
> *trans* (perpendicular to the UV light polarization).

The side chain polymers used by Shibaev *et al* are also shown in figure 5.2. The increase of the molar percentage of the azobenzene moiety results in a higher order of the azo-dye fragments. The order parameter of the azo-dye side chain polymers cannot be very high. However, it can be considerably increased where annealing of the photo-oriented azo-dye side chain polymer is used above the glass transition temperature [21].

The mechanism shown in figure 5.2 has been further observed by Gibbons *et al* who dissolved the diazodiamine molecules in a polymer matrix of a silicone polyimide copolymer [3]. After spin-coating procedure the homogeneous LC cell was prepared by rubbing the polymer layer. Then the cell was illuminated by an Ar^+ laser near the wavelength of the maximum absorption of the dye ($\lambda \approx 490$ nm) with the direction of the illumination parallel to the rubbing axis (figure 5.3). Within the illuminated region, the azo-dye molecules were aligned perpendicular to the light polarization direction, thus resulting in a twist alignment of the LC cell [3]. After this, Gibbons *et al* observed the aligning effect of azo-dyes in LC volume, using a 'guest–host' mixture [22]. Later Gibbons *et al* used the LC

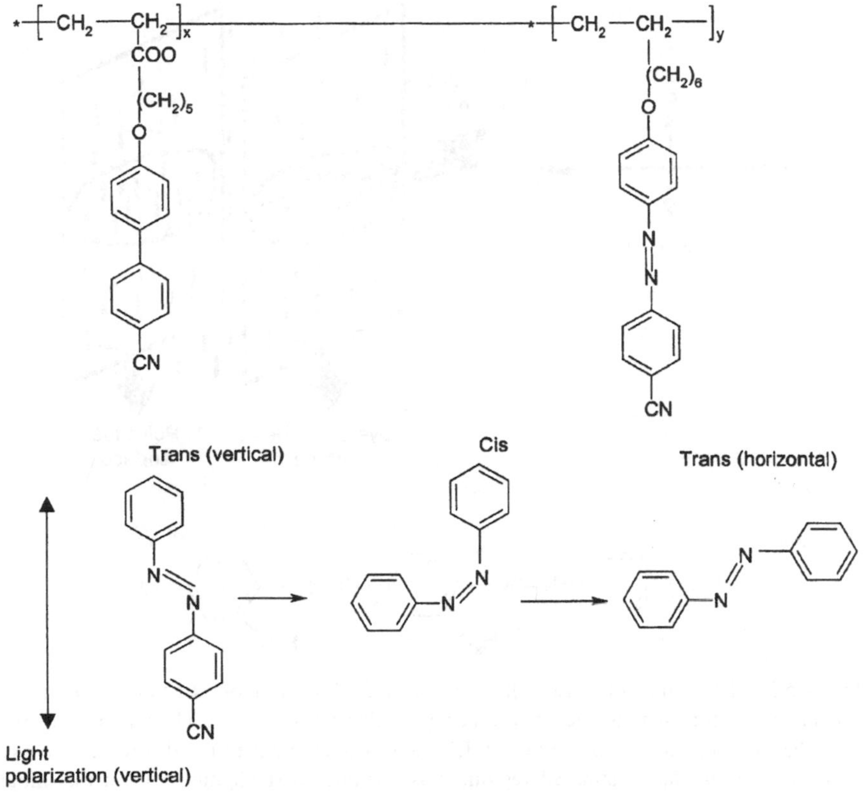

Figure 5.2. Side-chain polymers, used by Shibaev *et al* [19], showing the effect of *cis–trans* isomerization in the field of a polarized UV light. The *trans* isomer, which is parallel to the UV light polarization vector, is transferred to the *cis* isomer and then again to the *trans* isomer, which is perpendicular to the initial one. As a result of this transformation, all the absorption oscillators of azobenxene side-chain molecules align perpendicular to the UV light polarization.

cell of 40–50 µm thickness with initial homogeneous alignment on azo-dye/ polyimide coated substrate to record grey-scale images, produced by Ar$^+$ laser light with a wavelength of 514.5 nm [23, 24]. The reorientation of the azo-dye molecules perpenducular to the polarization of the UV light caused the subsequent reorientation of the neighbouring LC layer, thus increasing the contrast of the optical image.

5.2.2 Pure reorientation of the azo-dye chromophore molecules or azo-dye molecular solvates

Another mechanism is related to the reorientation of the dye molecules due to the action of the polarized light illumination. Besides chemical reaction,

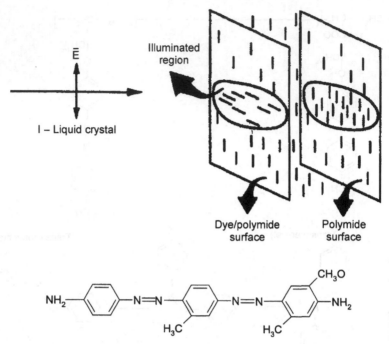

Figure 5.3. The aligning of an LC cell using the effect of rotation of the azo-dye absorption oscillator perpendicular to the activating light polarization vector. The structure of the used diazodiamine dye is shown below. LC molecules are aligned parallel to the azo-dye molecules within the illuminated region, thus forming twist alignment from the initial homogeneous configuration [3].

the UV light can induce an asymmetric potential field under which the stable configuration is characterized by the dye absorption oscillator perpendicular to the induced light polarization. One of the possible photo-aligning mechanisms in azo-dye films is a pure reorientation of azo-dye molecules [12, 25]. When the azo-dye molecules are optically pumped by a polarized light beam, the probability for the absorption is proportional to $\cos^2 \theta$, where θ is the angle between the absorption oscillator of the azo-dye molecules and the polarization direction of the light (figure 5.4). Therefore, the azo-dye molecules which have their absorption oscillators (chromophores) parallel to the light polarization will most probably get the increase in energy, which results in their reorientation from the initial position. This results in an excess of chromophores in a direction at which the absorption oscillator is perpendicular to the polarization of the light.

　　We have studied this photo-induced reorientation effect carefully. In our case the chromophore is parallel to the long molecular axis of the azo-dye (figure 5.4), i.e. azo-dye molecules tend to align their long axes perpendicular to the UV light polarization. The function $f(\theta)$ of the statistical distribution

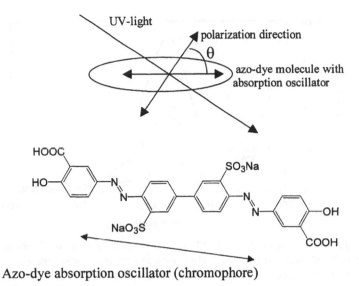

Azo-dye absorption oscillator (chromophore)

Figure 5.4. The qualitative interpretation of the photo-induced order in photochemical stable azo-dye films. Above: the geometry of the effect; below: azo-dye molecule, having the absorption oscillator (chromophore) perpendicular to the long molecular axis [12, 25].

of the azo-dye molecular aligning along the various orientations θ, which is $f = \frac{1}{4}\pi$ in the initial state, will tend to $f = \delta(\theta - \frac{1}{2}\pi)$ for a sufficiently long exposure. Hence, a thermodynamic equilibrium in the new oriented state will be established. Consequently, the anisotropic dichroism or birefringence is photo-induced permanently and the associated order parameter as a measure of this effect goes to the saturation value, which can be very large in these materials.

The reorientation of the azo-dye molecules perpendicular to the polar-ization of the polarization of the activated light was first observed by Kozenkov *et al* in Langmuir Blodgett (LB) films [26, 27]. The LB films, made of azo-dyes with a general formula shown in figure 5.5, were trans-ferred from the surface of water to the polished quartz glass. The induced value of optical anisotropy measured at the wavelength $\lambda = 632.8$ nm was $\Delta n \cong 0.23$. The dichroic ratio at the wavelength of the maximum absorption

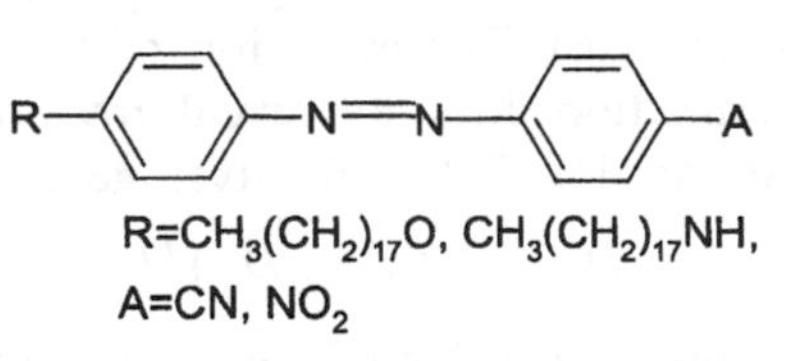

$$R\!-\!\!\text{⬡}\!-\!N\!=\!N\!-\!\text{⬡}\!-\!A$$

$R=CH_3(CH_2)_{17}O, CH_3(CH_2)_{17}NH,$
$A=CN, NO_2$

Figure 5.5. The azo-dyes, which are used as for the first demonstration of the effect of the reorientation of the dye absorption oscillator perpendicular to the polarization direction of the activating light through *cis–trans* isomerization [26, 27].

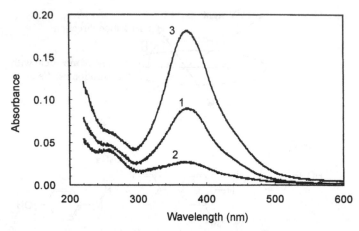

Figure 5.6. Absorption spectra of the SD-1 layer before the polarized UV exposure (curve 1). Curves 2 and 3 show the polarized absorption spectra after the exposure by a polarized UV light in the direction parallel ($D_{\parallel}$) and perpendicular ($D_{\perp}$) to the activating light polarization respectively.

of the dye $\lambda = 500$ nm was $N = D_{\perp}/D_{\parallel} \geq 4.6$, that corresponds to the order parameter $S = (1 - N)/(1 + 2N) = -0.35$ (70% of the dye molecules are arranged perpendicular to the light polarization vector [28, 29]).

In order to study the photo-reorientation of the azo-dye molecules (water-soluble sulfonic dye SD1) by the linearly polarized UV exposure, the polarized absorption spectra for the layer before and after the irradiation with the linearly polarized UV light were measured, using the incident light with polarization directions parallel and orthogonal to the polarization direction of the activated linearly polarized UV light [12, 25]. Figure 5.6 shows the polarized absorption spectra (absorbance or optical density) before (curve 1) and after (curves 2 and 3) the UV irradiation. Before the irradiation the absorption of the azo-dye layer does not depend on the polarization of the light, used in measurements. After the irradiation by linearly polarized UV light, the absorption of light with the polarization direction parallel to the polarization direction of the activated light ($D_{\parallel}$) decreases (curve 2, figure 5.6) while that with orthogonal polarization direction ($D_{\perp}$) increases (curve 3, figure 5.6). The evolution of the polarized absorption spectra after UV illumination does not reveal any noticeable contribution of photochemical reactions [12, 25], as the average absorption

$$D_{\text{ave}} = (D_{\parallel} + 2D_{\perp})/3 \qquad (5.1)$$

remains the same for any fixed value of the exposure time (figure 5.6). The order parameter S of the azo-dye chromophores can be expressed as [28, 29]

$$S = (D_{\parallel} - D_{\perp})/(D_{\parallel} + 2D_{\perp}), \qquad (5.2)$$

where $D_{\|}$ and $D_{\perp}$ are, respectively, absorption (optical density) of parallel and orthogonal polarized light to the polarization of the activated UV light. The order parameter S of SD1 is equal to -0.4 at $\lambda_m = 372$ nm (absorption maximum), which is 80% from its maximum absolute value $S_m = -0.5$ in our case. The dependence of the photo-induced birefringence δ, which is proportional to the order parameter S [28, 29], is

$$\delta = kS \qquad (5.3)$$

versus exposure time t_{exp}, measured to high accuracy with the help of the PEM (photo-elastic modulator) saturates for a sufficiently high value of t_{exp} [30]. The rate of the order parameter increase with the exposure time becomes higher for the higher values of the illumination power, which is in agreement with a model of Brownian rotatory diffusion of azo-dye molecules under the action of polarized light [30, 31] (figure 5.7).

5.2.3 Crosslinking in cinnamoyl side-chain polymers

The crosslinking of polyvinyl 4-methoxy-cinnamate (PVMC) under the action of linearly polarized light was first observed in 1977 [1], where the $(2 + 2)$ cycloaddition reaction of the properly located cinnamoyl fragments

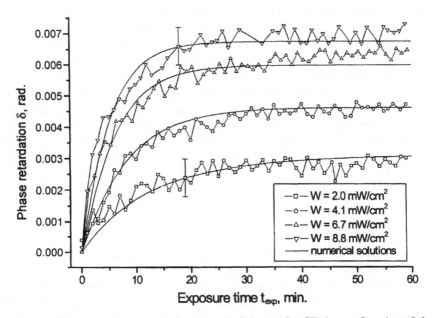

Figure 5.7. The photo-induced birefringence δ of the azodye SD-1, as a function of the exposure time for different exposure power. The experimental data are an agreement with the calculations according to the model of Brownian rotatory diffusion of azo-dye molecules under the action of polarized light [30].

was mentioned [32]. However, the first experiment of LC aligning by PVMC film, where the aligning material was clearly mentioned and the mechanism explained, appeared in 1992 [4]. The results of LC aligning by photo-polymerized PVMC films under the action of UV linearly polarized ($P_{\parallel}$) light ($\lambda = 320$ nm) were summarized as follows [4] (figure 5.8).

1. Linear photo-polymerization (LPP) leads to a preferred depletion of the cinnamic side-chain molecules along $P_{\parallel}$ due to the $(2+2)$ cycloaddition reaction.
2. This causes an anisotropic distribution of cyclobutane molecules with their long axis preferably aligned perpendicular to $P_{\parallel}$.
3. LC molecules are also aligned perpendicular to $P_{\parallel}$ (the axis of the preferred cyclobutane molecules alignment) due to the van der Waals (or dispersion) interaction.

Many publications concerning the mechanism of the photo-alignment in various cinnamate side-chain polymers, which provide a homogeneous LC alignment, were reviewed in [8]. According to some of them, the LC alignment is parallel to the polarization direction of the UV polarized light. The effect may take place both due to the photochemical *cis–trans* isomerization and dimerization process.

Coumarin side-chain polymers can be also cross-linked by $(2+2)$ cycloaddition reaction, and it is clear that *cis–trans* isomerization cannot occur [8]. The dimerized product is much greater in this case and LC alignment is parallel to the polarization of the activating UV light [8, 34] (figure 5.9). The superior properties of these materials in comparison with cinnamate side-chain polymers were first reported by Schadt *et al* [35].

Other crosslinking materials suitable for photo-aligning applications were mentioned elsewhere, in particular, side-chain polymer liquid crystal (SPLC) with cinnamoyl and biphenyl fragments [36], low-molecular-weight photo-cross-linkable composites (LMWPC) [37], polysiloxane cinnamate side-chain polymers [38] and others [39, 40]. Certain photo-aligned side-chain polymer materials obtained by the crosslinking effect are shown in reference [8].

5.2.4 Photodegradation in polyimide materials

As polyimide (PI) materials possess high temperature stability and are commonly used in the LCD industry, it was highly desirable to modify them for the photo-alignment applications. The first results in this field show that the polarized light with $\lambda = 257$ nm can induce LC alignment perpendicular to the direction of polarized UV light [6]. Thus the direction of the maximum density of unbroken polyimide chains, which is perpendicular to the light polarization, defines the direction of LC alignment. The direction of the photo-alignment can be changed by varying the direction

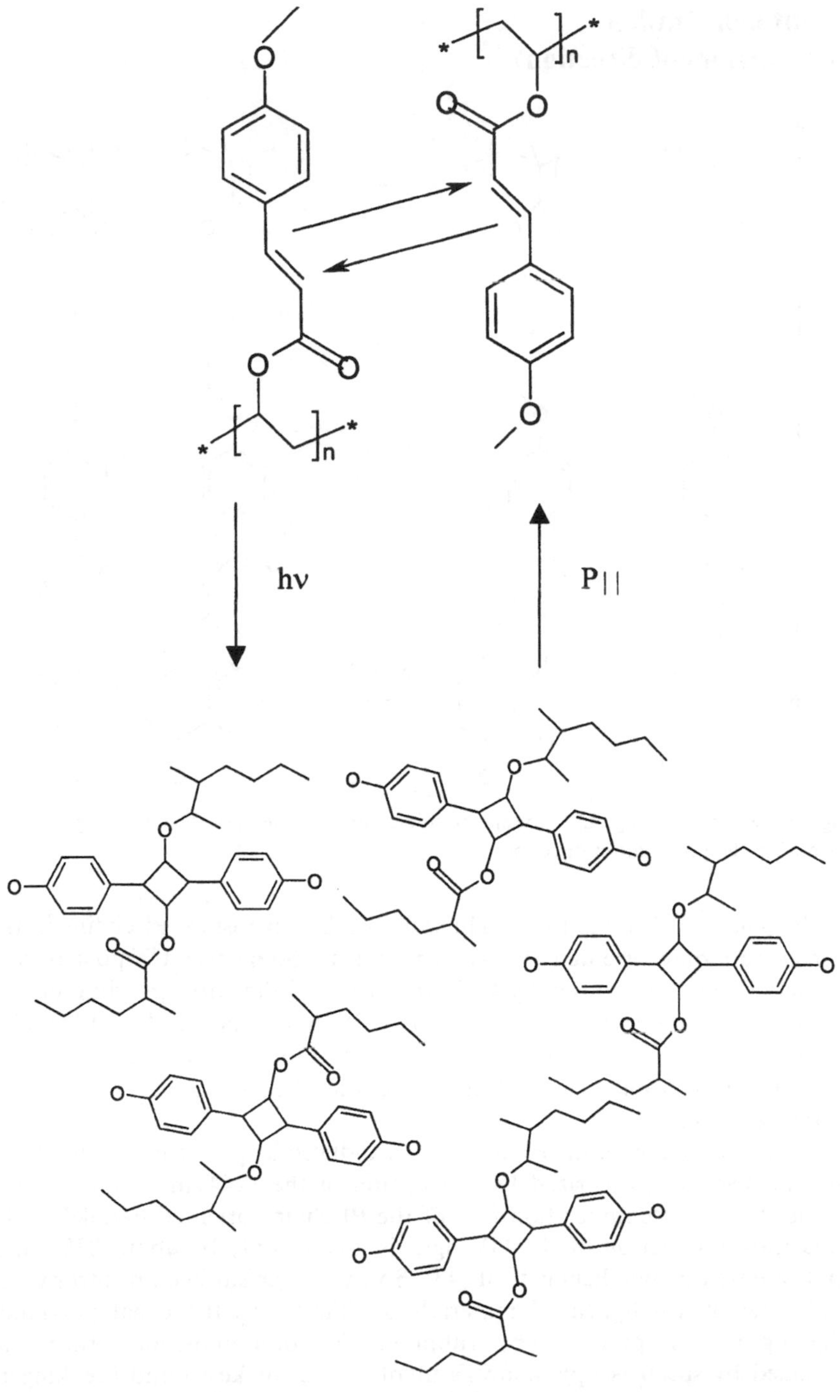

Figure 5.8. The mechanism of crosslinking in polyvinyl 4-methoxy-cinnamate [4].

Light polarization
(LC alignment direction)

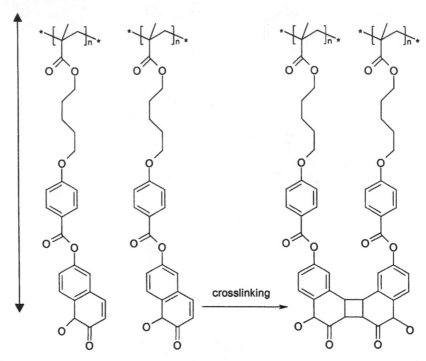

Figure 5.9. Crosslinking in a coumarin side-chain polymer [8]. The LC alignment is parallel to the polarization direction.

of the polarized UV light [41]. Thus, before UV exposure, PI chains in the film are randomly oriented. PI chains parallel to the exposed UV polarization are selectively decomposed by UV exposure, and the corresponding photoproducts become randomly relocated in PI film. The residual PI chains perpendicular to the exposed UV polarization remain unchanged and cause the anisotropic van der Waals forces to align LC molecules along its optical axis [42].

The anisotropy of the PI film surface induced by polarized UV light was investigated using polarized IR absorption of the $1244\,\mathrm{cm}^{-1}$ band. It was found that the decomposition rate of the PI chains oriented parallel to the polarization direction of the UV light is greater only by about 23% than that oriented perpendicular to it [43]. So the larger surface anisotropy can make the photo-aligned PI materials as efficient as the usual polyimide aligning agents, prepared by rubbing. The decomposition process as obtained by spectroscopic analysis involves the weakest bond breaking to form free radicals, which are responsible for the oxidation reaction [44].

There are several drawbacks of the LC photo-aligning materials, prepared by photochemical reaction and the photo-degradation in particular, as follows.

1. A small value of the order parameter (2) and the corresponding low value of the induced optical anisotropy and dichroism [43].
2. The order parameter is very sensitive to the exposure time and chemical content of the substance and has to be accurately controlled. For sufficiently long exposure times the order parameter comes through the

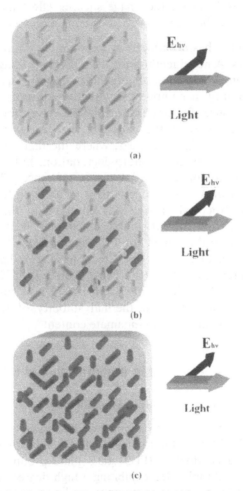

Figure 5.10. Photochemical reaction in the photosensitive layer where the molecular motion is hindered. After a sufficiently high exposure time the polarization of the activated light $E_{h\nu}$ will induce the photochemical transformation not only in the molecular units having the absorption oscillator parallel to the polarization, but within the whole molecular ensemble of the photosensitive layer [46].

maximum and goes to zero [45]. The latter was first mentioned in the early work of Kozenkov *et al* [46], who predicted this effect for the photo-sensitive layers, where the molecular movement is hindered. Indeed, for sufficiently long exposure time the photo-induced transformation will take place not only for the molecular units, in which the absorption oscillator forms a small angle of the activating light polarization, but for all the molecular ensemble, thus leading to a zero order (figure 5.10).
3. The contamination of the initial substance by the by-products of the photo-degradation is possible in certain cases. The by-products can produce ions and initiate the image sticking effect and flicker [47], as well as reduce the thermal stability of the LC alignment [48].

Certain research has been made to improve the photo-aligning performance of PI layers. A novel method for LC alignment using *in-situ* exposure of the PI film during imidization of polyimide resulted in a sufficient increase of the thermostability in comparison with the conventional method that employed UV exposure after imidization [49]. However, the thermal stability deteriorated considerably if the exposure time became too long. New materials based on polyimide appear, where the main mechanism of the LC photo-alignment was not a photo-degradation, but the orientation of the main chain of the polymer was shown in reference [50]. The laser light with a wavelength of $\lambda = 355$ nm was used to activate the polyimide film and no obvious structural change was observed in this case. The new photo-crosslinkable and solvent-soluble polyimides, containing cinnamate side-chains, were also developed for LC aligning [51]. The LC alignment was obtained perpendicular to the direction of UV light ($\lambda = 350$ nm) polarization and remains intact after being heated at 85 °C for 450 h. New polyimides with pendant cinnamate groups were also suitable for photo-aligning [52]. However, the best thermal stability of the photo-aligned layer was observed at the lowest cinnamate content.

5.3   LC surface interaction in a photo-aligned cell

5.3.1   Improvement of materials for photo-aligning

The first agents used for LC photo-aligning were polyvinyl-cinnamate (PVCN) derivatives [4, 5, 53–55]. Photo-aligned polyvinyl 4-methoxy-cinnamate (PVMC) film on a conducting (ITO) glass plate produced a well-ordered array of liquid crystal molecules, exhibiting a high degree of orientation and positional order, as seen by the scanning tunnelling microscopy (STM) technique [53]. A uniform pretilt angle can be also generated [55, 56]. However, the measured azimuthal anchoring energy (4.2×10^{-6} J/m^2) [54] was considerably lower than that for the rubbed polymers, and the degradation of the LC alignment was observed at the temperatures exceeding 70 °C [57, 58]. This was the reason for the development of the new cross-linkable cyanobiphenyl cinnamate derivatives, which exhibit temperature-stable photo-alignment, especially after coating by a polymer film. The new alignment materials were called polymeric liquid crystals–linear photo-polymerized (LCP-LPP) films and proved to be useful not as the aligning layers, but as integrated optically patterned retarders and colour filters [57]. A number of new LCP-LPP photo-aligning materials were recently announced for liquid crystal on silicon (LCOS) applications [59]. The voltage holding ratio (VHR) was found to be even higher than for standard rubbed PI layers even after heating at 120 °C for 50 h. Thus active matrix addressing of LCOS became suitable as the effect of ionic charge was minimized. The sufficiently high pretilt angle of 10° was obtained on both reflective and transmissive substrates of an LC device at a sufficiently large LPP thickness of 200 nm.

A perfect quality homogeneous alignment was obtained also on polyimide films, but a relatively high energy of illumination ($\sim 7\,\mathrm{J/cm^2}$) was needed to get the required maximum order parameter of PI film [6]. If the illumination energy ($\lambda = 257\,\mathrm{nm}$) exceeded $7\,\mathrm{J/cm^2}$, the alignment became loose and soon the mono-domain LC alignment was not induced in accordance with the origin of the photo-chemical mechanism of photo-alignment (figure 5.10). The progress of photo-aligning materials based on photosensitive PI layers was reported by Gibbons *et al* [60]. The polyimide host mixture comprises special types of the molecular dopants, which promote the optimal photo-response, induce pretilt angle, sufficiently high anchoring energy, better dielectric properties and adjust the material for commonly-used offset printing technology for LCD production [29]. The equipment development for LC photo-aligning was also demonstrated.

The azo-dyes for LC photo-aligning were investigated [12, 25, 61]. The photo-aligning of azo-dye takes place due to the pure reorientation of the molecular absorption oscillators perpendicular to the UV light polarization, as discussed above (figure 5.4). The temperature-stable pretilt angle of 5.3° was obtained by a two-step exposure of azo-dye film using normally incident polarized light followed by oblique non-polarized light. The azimuthal anchoring energy of a photo-aligned substrate was $W_\varphi \approx 10^{-4}\,\mathrm{J/m^2}$, the same as the anchoring of the rubbed polyimide (PI) layer. The value of voltage holding ratio (VHR) measured for the photo-aligned LC cell was found to be even higher than for rubbed PI layer, which enables the azo-dye applications as aligning layers in active matrix liquid crystal displays (AM-LCDs). The thermal stability of the photo-aligned azo-dye layers was sufficiently high, while UV-stability was also improved by polymerization.

5.3.2 Pretilt angle generation in photo-aligning materials

The value of the pretilt angle of an LC cell on the substrate is one of the important characteristics for photo-alignment. In most of the electro-optical

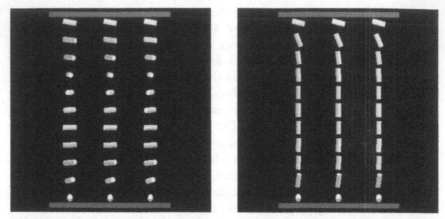

Figure 5.11. LC configuration in an STN-LC cell with pretilt angle on the boundaries, when the applied voltage U is switched from $U = U_{\text{off}}$ (left image) to $U = U_{\text{on}}$ (right image). If the pretilt angle at the boundary is not high enough, the switching between left and right configuration is not possible due to the hysteresis effect [28, 29].

modes the pretilt angle should be sufficiently high to enable a quality performance of LCD [29]. For instance, in supertwisted nematic LCD (STN-LCD) the value of the pretilt angle at the substrates should be high enough to provide a perfect switching between the supertwisted and almost homeotropic configurations to avoid the hysteresis phenomena (figure 5.11) [28, 29].

It is easy to understand that, if the pretilt angle is not defined, when normally incident polarized light activated the molecules in the photo-sensitive layer, the two molecular symmetrical pretilt angles shown in figure 5.12 are

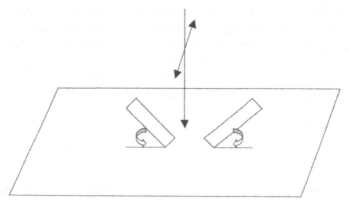

Figure 5.12. The pretilt angle is not defined, when normally incident polarized light activated the molecules in the photo-sensitive layer. The two molecular symmetrical pretilt angles shown are both possible, because the absorption oscillators are perpendicular to the activated light polarization in both cases. The molecules, similar to those described in figure 5.5, are shown by rods.

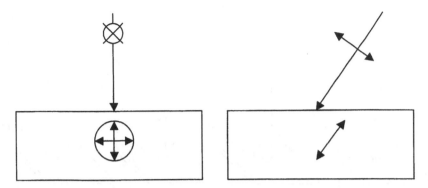

Figure 5.13. Two-step exposure by polarized light to induce the pretilt angle in the photo-aligned film [63]. After the first exposure the molecules are arranged in one plane perpendicular to the light polarization. The second exposure chooses the alignment perpendicular to the direction of p-polarization of the obliquely incident light (parallel to the light propagation).

both possible, because the absorption oscillators are perpendicular to the activated light polarization in both cases. Thus, to induce the pretilt angle, the symmetry of the case should be broken by changing either the direction of light incidence or the procedure of the LC cell preparation. One possibility to remove the degeneracy of the pretilt alignment direction is to involve the direction of LC flow, when the cell is filled with LC [62]. However, in this case the subsequent heating and cooling of the LC cell to induce LC ⇔ isotropic liquid transition will again restore the degeneration in the pretilt angle, as the filling direction will be 'forgotten'.

The method of double polarized UV exposure with different polarization and irradiation angles was first proposed by Iimura *et al* (figure 5.13) [63]. The normally incident polarized light first changes the random direction of the molecules to the alignment perpendicular to the polarization, i.e. in one plane, and after this the obliquely incident light with a p polarization chooses the alignment direction perpendicular to the polarization or parallel to the propagation direction. However, the pretilt angles obtained by Iimura *et al* were rather small and did not exceed 0.3° [63]. The effect can be partially explained by the fact that the really observed LC pretilt angle is dependent not only on the direction of the anisotropic order in the photo-aligning layer, but also on the properties of the LC mixture, which is in contact with this layer. In other words the final result will be dependent on the interaction of the two order parameters; photo-aligning layer and LC layer [29].

The pretilt angles can be also obtained in one-step exposure by obliquely incident polarized light, in the case where the aligning is induced parallel to the polarization direction [64]. Any desired LC pretilt angles from 0° to 90°

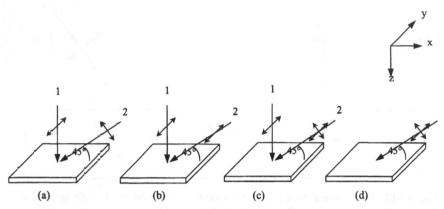

Figure 5.14. Various methods of UV light irradiation to attain the pretilt angle of the LC layer on the photo-aligned film. Case (a): two step exposure by y polarized light vertical to the plane of the azo-dye film (x–y plane) and oblique p polarized light; (b): case (a) with s polarized light used at the second step; (c): case (a) with non-polarized light at the second step; (d): one-step exposure with oblique non-polarized light. The activated UV light propagates in the x–z plane at an oblique angle with the substrate $\theta = 45°$ [61].

have been demonstrated, using polyimide films with different molar fraction of fluorine fragments. However, the value of the pretilt angle drops down for the sufficiently long exposure time, which is the consequence of the photo-chemical reaction in PI discussed above (figure 5.10). The pretilt angles can be induced by one-step exposure of polarized light, when coumarin-substituted cinnamate derivatives were used (figure 5.9) [35]. Pretilt angles of about 3° were produced, when polyimide film was illuminated by non-polarized obliquely incident UV light [64, 65]. Certain researches were made to generate pretilt angles combining the homogeneous alignment, produced by UV illuminated PI films, doped with surfactants, such as lecithine [66]. However, only the angles close to homogeneous alignment (1.8°) or homeotropic alignment (89.5°) can be produced depending on the concentration of lecithine.

Four kinds of exposure method were tested to produce the pretilt angle, using azo-dye materials (figure 5.14) [61]. To measure the pretilt angle, the LC cells were assembled with substrates antiparallel to UV irradiation directions and were filled with an LC mixture. The crystal rotation method was used to measure their pretilt angles [29]. Pretilt angles of 0.7°, 0.1°, 5.3° and 3.8° are obtained with the exposure method (a), (b), (c) and (d), respectively (figure 5.14). The oblique irradiation of non-polarized light proved to be the most efficient in order to obtain the larger pretilt angles. The dependence of the pretilt angle on the exposure energy in method (c) is shown in figure 5.15. The pretilt angle was temperature-stable and did not change after heating the sample up to 100 °C for 10 min.

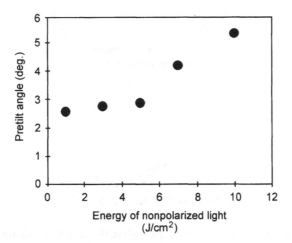

Figure 5.15. Dependence of the pretilt angle on the energy density of the obliquely irradiated nonpolarized light after the irradiation of vertical polarized light (figure 5.14(c)) with an energy density of $10 \, \text{J/cm}^2$ [61].

5.3.3 Anchoring energy in photo-aligning materials

The anchoring energy is one of the most important parameters which characterize the LC alignment quality [28, 29]. It usually consists of two parts, azimuthal and polar, that describe how easy it is to change the alignment of the LC director at the surface. The anchoring energy is a part of the LC free energy, needed to realign the LC director from the preferred orientation at the substrate, e.g. z axis ($\theta = \varphi = 0$, figure 5.16) up to the angles θ and φ in polar and azimuthal direction [28, 29]:

$$W(\theta, \varphi) = \tfrac{1}{2} W_\theta \sin^2 \theta + \tfrac{1}{2} W_\varphi \sin^2 \varphi \approx \tfrac{1}{2} W_\theta \theta^2 + \tfrac{1}{2} W_\varphi \varphi^2 \qquad (5.3)$$

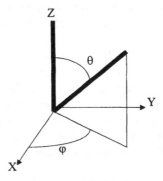

Figure 5.16. The anchoring energy is a part of the LC free energy, needed to realign the LC director from the preferred orientation at the substrate, such as e.g. z axis ($\theta = \varphi = 0$) up to the angles θ and φ in polar and azimuthal direction: $W(\theta, \varphi) = W_\theta \theta^2 + W_\varphi \varphi^2$, where W_θ and W_φ are the corresponding polar and anchoring energy [28, 29].

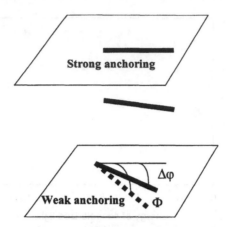

Figure 5.17. The value of the apparent twist angle in the LC cell $\Delta\varphi$ is lower than the angle Φ set by the preferred azimuthal director alignment on the substrates (shown by solid and dashed lines respectively), because the strong azimuthal anchoring substrate affects the director alignment on the weak azimuthal anchoring one [70, 71].

where W_θ and W_φ are the corresponding polar and anchoring energy. The value of W_θ and W_φ is a very important characteristic of LC–surface interaction.

To avoid the formation of surface walls and to provide a fast 'off' switching time, the anchoring energy should be sufficiently high, e.g. $W_\varphi > 10^{-4}\text{J/m}^2$, comparable with that of the rubbed polymer surface. This is not really observed for poly-vinylcinnamate (PVCN) derivatives [54]. Recent measurements also confirm that $W_\varphi \approx 10^{-5}$–$10^{-6}\,\text{J/m}^2$, i.e. about one or two orders of magnitude smaller than the required value [67, 68]. The polar anchoring strength, measured for the polar monolayers, was an order of magnitude higher [69]: $W_\theta \approx 10^{-4}\,\text{J/m}^2$.

The azimuthal anchoring energy can be measured using a substrate with a rubbed polyimide alignment layer, and that with a photo-aligned layer assembled with a twisted nematic configuration (figure 5.17) [70, 71]. The value of the apparent twist angle in the LC cell is lower than that set by the preferred azimuthal director alignment on the substrates (shown in figure 5.17 by solid and dashed lines respectively), because the strong azimuthal anchoring substrate affects the director alignment on the weak azimuthal anchoring one. The azimuthal anchoring energy of the photo-aligning layer W_φ can be calculated from the torque balance equation in the LC cell, provided that the anchoring energy of the rubbed surface is infinite [70, 71]:

$$W_\varphi = 2K_{22}\Delta\varphi/d\sin[2(\Phi - \Delta\varphi)], \qquad (5.4)$$

where Φ, $\Delta\varphi$, K_{22} and d are the twist angle set by the preferred alignment directions on the two LC cell substrates, apparent twist angle, twist elastic constant of liquid crystal, and cell gap, respectively. In [61] the measured apparent twist angle $\Delta\varphi$ was equal to $\Phi = 80°$ with an accuracy of $1°$, for the whole range of the UV light exposure energies between $1\,\text{J/cm}^2$ and $10\,\text{J/cm}^2$, which states that the anchoring energy $W_\varphi > 10^{-4}\,\text{J/m}^2$, which is comparable with the anchoring energy of the rubbed polyimide layer [29].

5.4 Applications

5.4.1 Multi-domain LC cells

Multi-domain configuration is suggested to be one of the most important to obtain the wide viewing angles of LCD (figure 5.18) [29]. The idea is simple: each twisted nematic (TN) LC cell has the best viewing angles only at the quarter of the whole 2π azimuthal angle, so we have to combine four different TN configurations in one pixel of LCD (figure 5.18). The practical realization of multi-domain samples, based on a multi-step photolithographic process [72], preparation of special polymer layers with two randomly distributed pretilt angles [73] or rubbing only one substrate with a proper doping of the LC with a chiral additive [74] are complicated, so photo-patterning remains one of the effective techniques to produce a multi-domain LCD cell [75]. We have to mention here that multi-domain LC cells with almost homeotropic alignment (vertical aligned nematic (VAN) mode) can be effectively made by the preparation of a special dielectric pattern (protrusions) on one of the electrodes of the LC cell [76]. A new method of fabricating VAN-LCD, using multi-domain alignment with self-aligned four-domain on polymer grating, was also proposed recently [77, 78].

The production of four-domain TN LC cells has been reported, using two photo-aligned polyimide substrates with stripes of opposite pretilt angle [79] and polysiloxane photopolymer with a controllable pretilt angle [80]. A first $10.4''$ active matrix addressed thin-film transistor (TFT-LCD) panel, using multi-domain configuration with symmetric and wide viewing angle was fabricated [81]. Later the multi-domain STN-LCD ($\varphi = 240°$) cells with wide viewing angles was demonstrated [82].

Certain success has been obtained in the development of multi-domain LCD, utilizing homeotropic orientation. A four-domain hybrid twisted nematic (HTN) LC cell was fabricated with a small pretilt on one substrate and high on the other, showing uniform transmission characteristics for wide viewing angles up to $40°$ [83]. Two-domain [84] and four-domain [85] photo-aligned VAN-LCDs were also made. The LC director was homeotropic on one substrate, while the other substrate was prepared with reverse bias tilt angles $\theta = \pm 89°$ causing the central director to tilt in opposite directions upon application of the voltage [84]. The dual-domain configuration (figure 5.19) was also used to prepare a $5''$ 320×240 TFT-TN-LCD with wide viewing

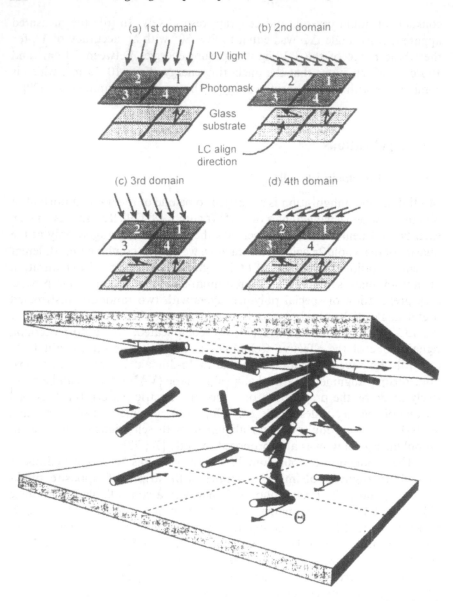

Figure 5.18. Multidomain orientation in a liquid crystal cell: (below) four-domain TN configuration, (above) application of a multi-step photo-aligning technique for its realization [29].

angles [86]. The concept of an LCD master, i.e. a micro-structured polarizer, which generates polarized light from non-polarized incident light with locally different polarization directions, was proposed to generate a multi-domain photo-aligned substrate for LCD cell by one-step exposure [85]. A passive

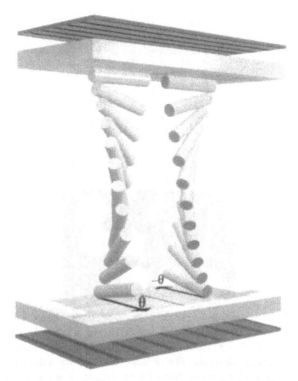

Figure 5.19. Dual-domain TN-LCD, prepared by photo-aligning [86].

multi-domain LC layer, which locally rotates the plane of the incident light polarization, was used for the purpose. A dual-domain 80° photo-aligned TN-LCD was prepared using an LCD master. A similar micro-structured LC polarizer has been mentioned also [87].

5.4.2 Photo-patterned phase retarders and colour filters

The effect of photo-patterning in poly(vinyl-cinnamate) derivatives (PVCN) is consequent to the phenomena of the photo-induced optical anisotropy, found during the illumination of PVCN film by polarized UV light within its self-absorption band [1]. The main properties of the PVCN materials are:

  (i) a comparatively high value of the photo-induced birefringence (PB) (more than 0.02 at a wavelength of 632.8 nm);
 (ii) partial reproducibility of the 'writing–erasing' cycles;
(iii) long-term memory of the recorded PB value (more than 25 years);
(iv) high spatial resolution (more than 8000 mm^{-1}).

First applications of the effect of photo-induced optical anisotropy in PVCN films comprise: writing the polarization holograms and images [88, 89],

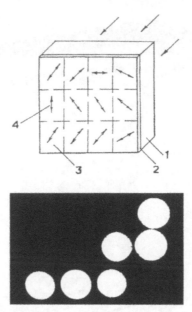

Figure 5.20. Multi-functional polarizer (above), comprising a linear polarizer (1) and a photo-patterned phase retarder (2) and its image (below) in crossed polarizers [91]. UV-irradiated (He-Cd laser, $\lambda = 325$ nm) PVCN film (2), used for the photo-patterned phase retarder has a thickness of 10–40 μm. The variation of the optical axis of the phase retarder (3, 4) was made by changing the direction of the polarization of the activated UV-illumination.

planar waveguide optics [90], polarization optical elements [91], optical data processing [92], non-destructive testing of defects, protection of valuable papers, as well as other devices by marking them with hidden labels, etc. [93]. Most of the early papers devoted to the above-mentioned applications were published in 1977–1987 in the USSR only, and are not well known worldwide. The photo-patterned polarization optical elements obtained by the exposure of PVCN films are shown in figures 5.20 and 5.21 [91].

The new impulse of the application of photosensitive media appeared when PVCN was discovered as an LC photo-aligning substance [4, 5]. The concept of the design of photo-patternable hybrid linear photo-polymerizable (HLPP) configuration was proposed by Schadt *et al* (figure 5.22) [57, 58]. The HLPP layers include a linearly photo-polymerized polymer (LPP) as the aligning layer and an LPP-LCP phase retarder in one compact implementation (figure 5.22). This technology opens a new way of constructing the new types of colour STN-LCDs, where the basic red, green and blue colours are created by specially adjusted phase retarders to each pixel of the LCD. The angle of the phase retarders with respect to the LC alignment and the corresponding value of the phase redardation can be calculated according to STN-LCD or TN-LCD configuration.

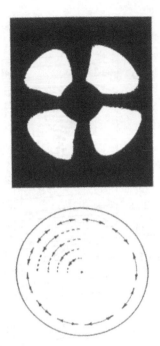

Figure 5.21. Photo-patterned phase plate (above) and its image (below) in crossed polarizers [91]. The PVCN plate was rotated at 2 rpm. The activated polarized UV light comes along the axis of the plate rotation. The exposure time was 5 min at the power of 10 mW/cm². The optical axis was induced in a circular direction and the value of the photo-induced birefringence (PB) linearly decreases in a radial direction.

Figure 5.23 shows one of the configurations of TN-LCD, which makes it possible to obtain birefringent colours, using a photo-patterned phase retarder. The colour is made using two phase retarders, PR1 and PR2, placed at symmetrical angles α with respect to the input polarizer. TN-LCD (or any other LCD configuration) rotates the 'horizontal' polarization after polarizer P_1 up to the angle of 90° and then the polarization comes through the 'vertical' polarizer P_2. It can be easily shown that if light passes through the double phase retarder construction (DPR) with the two symmetrically placed phase retarders PR1 and PR2 with respect to polarizer P_1 following by the output 'horizontal' polarizer P_3, then the output intensity I is written as:

$$ I = \tfrac{3}{8} I_0 \sin^2 4\alpha \sin^2 \pi \, \Delta nd / \lambda, \qquad (5.5) $$

where I_0 is the light intensity after the input polarizer P_1, Δnd is the phase retardation introduced by each of the two symmetrically placed retarders in DPR configuration (figure 5.23), and λ is the light wavelength. Angle α should be equal to 22.5° to provide the maximum transmission of the

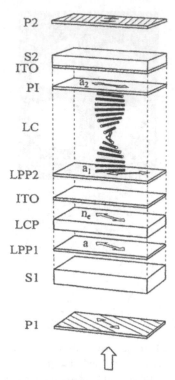

Figure 5.22. Hybrid linearly photopolymerizable (HLPP) configuration, used for STN-LCD (bottom substrate) [57]. P1, P2: polarizers; LPP1, LPP2: photo-aligning layers (linear photo-polymers); LCP: phase retarder (LC polymer); S1, S2: glass substrates; PI: rubbed polyimide layer; ITO: transparent electrodes; a, a_1, a_2: preferred directions of alignment; n_e: optical axis of the phase retarder.

light. The appropriate phase retardations Δnd for red, green and blue light are obtained by setting $\sin^2 \pi \Delta nd/\lambda = 1$, i.e. $(\Delta nd)_{\text{red}} = 315\,\text{nm}$ ($\lambda_{\text{red}} = 630\,\text{nm}$), $(\Delta nd)_{\text{green}} = 275\,\text{nm}$ ($\lambda_{\text{green}} = 550\,\text{nm}$), and $(\Delta nd)_{\text{blue}} = 240\,\text{nm}$ ($\lambda_{\text{blue}} = 480\,\text{nm}$). The advantage of the birefringent colour TN-LCD is low power consumption due to the high transmission in the 'off' state. The 'on' state produced by almost homeotropic LCD between crossed polarizers P_1 and P_2 is perfectly dark. The disadvantage of this structure is the angle dependence of the birefringent colours.

To produce wide viewing angles the retarder structure should be more complex, as suggested by Schadt *et al* [95]. Thus two orthogonal positive uniaxial plates behave as a negative phase plate [96]. The plates can be made on the basis of LPP-LCP technology, proposed by the Rolic Company [97]. Photo-patterned non-absorbing and bright interference colour filters, as well as positive and negative phase retarders based on photo-aligning technology, are presented in reference [96]. The stack of three TN-LCDs,

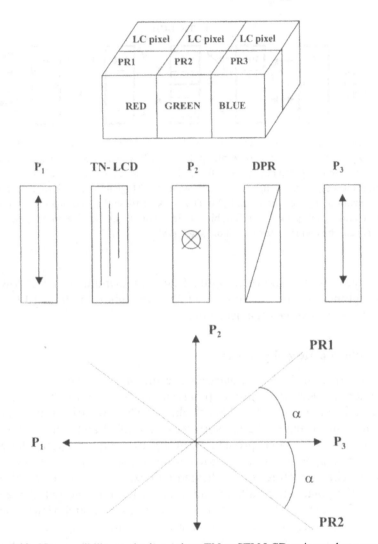

Figure 5.23. New possibility to obtain a colour TN or STN LCD, using a photo-patterned retarder [94]. Above: basic red, green and blue colours obtained by the corresponding properly adjusted phase retarders (PR1, PR2 and PR3). Below: one of the possible configurations with TN LCD, three polarizers ($P_1 \| P_3$) $\perp P_3$ and double phase retarder configuration (DPR) composed of the two phase retarders PR1 and PR2, placed at symmetrical angles α with respect to the input (output) polarizer $P_1 \| P_3$.

accompanied by yellow, magenta and cyan photo-patterned phase retarders respectively in a front and a back surface, can create a full colour LCD, whose transmission is about six times higher than that of conventional LCD with absorptive colour filters [98]. The typical configuration of

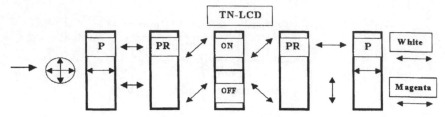

Figure 5.24. A stack photo-patterned magenta phase retarders (PR) in a combination with TN-LCD placed between two horizontal polarizers (P), which affects only green light and switches between magenta and white [98]. TN-LCD rotates the state of polarization in the 'on' state and keeps it unchanged in the 'off' state [29]. Magenta PR stack rotates up to 45° the polarization angle of the green light. The stack of these three configurations (yellow, magenta and cyan) provides a full birefringent colour LCD. The input and output polarizers can be common for the whole stack [98].

the magenta part is shown in figure 5.24. The manufacturing of low-cost high-brightness birefringent colour LCDs using photo-patterned LCP-LPP phase retarders can be organized [98].

5.4.3 Photo-patterned polarizers

Light-polarization films or polarizers are the major components of LCDs and other LC devices. Common polarizers are based on polyvinyl-alcohol (PVA) iodine films of 150–400 μm thickness. These polarizers are generally laminated on the external glass surfaces of the LCD and consist of a stack of thin films for scratch protection, anti-glare, anti-reflection, phase and chromatic compensations, anisotropic absorption, etc. [29]. The fabrication of this stacked structure is complicated and expensive. The external place-ment of the polarizers results in additional reflections and parallax effects, which affect the LCD contrast, optical performance and viewing angles [29]. So, thin internal polarizers (figure 5.25) are highly desirable. Yet, this can hardly be achieved on the basis of conventional polarizing film technology.

Amphiphilic dye molecules such as sulfonated chromogens can be used in place of the PVA-iodine for the preparation of thin polarizing films (figure 5.26) [99–104]. These molecules are usually water-soluble and, at a wide range of concentrations, temperatures and pH-values, they will self-assemble and stack up to form hexagonal complexes. This is usually referred to as the lyotropic liquid crystalline phase [28, 29]. By mechanical shear stress, an orientation of the lyotropic liquid crystal (LLC) is induced along the shear flow direction (figure 5.26(b)). When the solvent has evaporated, the molecular order is maintained in the LLC solid film. As a result, it gives rise to a high-order parameter and a cylindrical symmetry that along the

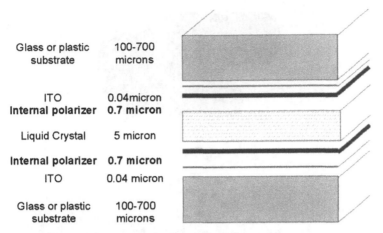

Glass or plastic substrate	100-700 microns
ITO	0.04micron
Internal polarizer	**0.7 micron**
Liquid Crystal	5 micron
Internal polarizer	**0.7 micron**
ITO	0.04 micron
Glass or plastic substrate	100-700 microns

Figure 5.25. Superthin internal polarizer for LCD [99].

c axis of these complexes most of the e-wave is transmitted, whereas the o-wave, which propagates on the plane orthogonal to this c axis, can be absorbed effectively. These dye molecules are also stable against the UV radiation, thermal treatment (up to 230 °C), and have good colour fastness [99–102]. These properties make them attractive for the replacement of iodine-based external polarizers in LCD production.

However, there are problems associated with the shear-flow induced alignment of the LLC. The first is the visual defects, which are in the form of horizontal stripes of several to tens of microns wide. The defects divide the area with different molecular orientations, which are clearly seen in a polarized light as vertical bands. These are due to the non-uniform LLC flow gradients, which are sensitive to the shear speed, LLC viscosity and temperature. The second problem is the particulate contamination, which is transferred from the substrate to the rollers and back onto another substrate. This is a serious problem since the stringent particle control and frequent cleaning will otherwise slow down the throughput and add a new cost. In addition, the polarization axis cannot be made arbitrary to follow a specific local distribution, for example, to follow the mosaic picture with the characteristic size of tens of microns or less. This limitation is due to the poor spatial resolution of this method.

The fabrication methods of the e-wave polarizer, based on the photo-alignment method, was first proposed by Kozenkov *et al* [103]. The polarizing direction in the film is determined by the polarization vector and the incidence plane of the actinic radiation. Since the local structure is patterned optically, the multi-axes and multi-colour polarizers can be prepared cost-effectively. In addition, the photo-aligning is a non-contact method, which minimizes the particulate contamination and defect

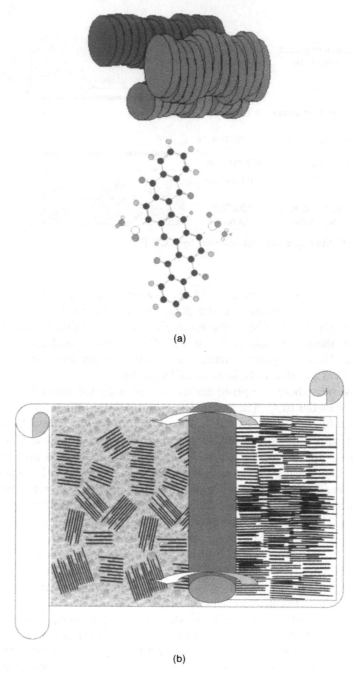

(a)

(b)

Figure 5.26. (a) Typical structure of amphiphilic dye molecules, stacked to columns to form lyotropic LC crystal. (b) Alignment of lyotropic LC by mechanical flow to form a polarizing layer [99].

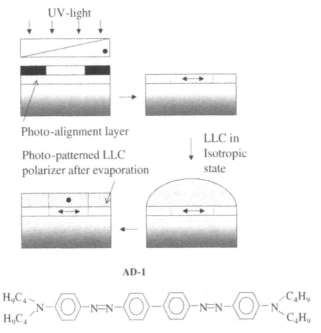

Figure 5.27. Method of fabrication of the e-wave polarizer [103]. The photo-aligning layer is prepared based on the photo-aligned AD-1 layer with a chemical formula, shown below. LLC is an isotropic phase put on top of the photo-patterned AD-1 layer. The polarizer was formed after the evaporation of the solvent. The absorption axis of the polarizer was parallel to the direction of AD-1 molecules (perpendicular to the direction of the polarized UV light).

generations. This polarizing film, which is about 0.3–0.7 μm, can be placed on the internal or external substrate surfaces of an LCD (figure 5.25). For the TN-LCD with internal polarizers, the electro-optic characteristics were basically similar to those with the external polarizers [104]. This is central to internal polarizer development since the STN-LCD is sensitive to any voltage and thickness variations.

The lyotropic liquid crystal Crystal Ink™ from Optiva, Inc. and azo-dye AD-1 was used for the photo-aligning experiments [103] (figure 5.27). The azo-dye order parameter (2), measured by dichroic spectra $S > 0.86$, while the dichroic ratio $N = D_\perp/D_\parallel > 20.4$ at $\lambda = 490$ nm. In the first fabrication stage, an azo-dye layer AD-1 of about 0.1 μm thickness is spin-coated on a glass substrate. It was then illuminated with a shadow mask using the polarized UV light to form the photo-alignment layer. In the second fabrication stage, a few drops of an isotropic LLC solution were dispersed onto the photo-alignment layer. When the solvent was evaporated, the LLC was oriented preferentially along the photo-induced

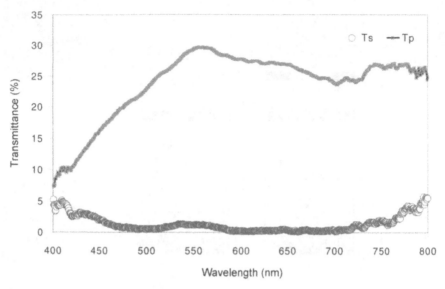

Figure 5.28. Polarization absorption spectra of the photo-aligned LLC polarizer. T_p and T_s represent the transmission spectra parallel and perpendicular to the direction of the polarized UV light, respectively [103, 104].

axis, where the absorption axis of the LLC was parallel to that of the photo-alignment layer in this case. The polarized transmission spectra of the lyotropic polarizer aligned by AD-1 are shown in figure 5.28. The polarization efficiency was calculated using

$$E_p = (T_p - T_s)/(T_p + T_s) \tag{5.6}$$

where T_p and T_s are the transmittance parallel and perpendicular to the direction of the polarized pumping light, respectively. At the peak transmission wavelength 559 nm, the polarization efficiency was equal to 92.2% and the transmittance was 29.7% [104]. The electro-optic characteristics of the TN-LCDs with the internal and external polarizers made on photo-aligned LLC were comparable and the results showed that the change in voltage and temporal responses were negligible. The picture (HKUST logo), obtained by photo-patterned LLC, viewed with a polarizer, having a polarization axis perpendicular to the absorption axis of the azo-dye AD-1 is shown in figure 5.29.

The photo-patterned colour (orange) polarizer can be prepared directly in a sufficiently thick (1 μm) layer of azo-dye AD-1 by the above-mentioned technique [103]. The transmission spectra of this alignment layer are shown in figure 5.30(a). The photoinduced anisotropy in this layer did not change for years and is stable up to 130 °C. The spectra can be made close

Figure 5.29. The picture (HKUST logo), obtained by photo-patterned LLC, viewed with a polarizer, having a polarization axis perpendicular to the absorption axis of the azo-dye AD-1 [104]. The exposure energy was $25\,\mathrm{J/cm^2}$ at $\lambda = 405\,\mathrm{nm}$.

to neutral by a special treatment with iodine or other vapours (figure 5.30(b)) [105].

5.4.4 Security applications

A new generation of optical security elements based on photo-patterning technology enables the fabrication of optical security devices. The photo-patterned security mark can be applied directly to documents via printing or coating methods as well as indirectly by hot-stamping processes (figure 5.31) [93, 97]. A latent image is created on the transparent substrate (e.g. plastic film, paper, glass, clock, product of motor-car or aircraft industry,

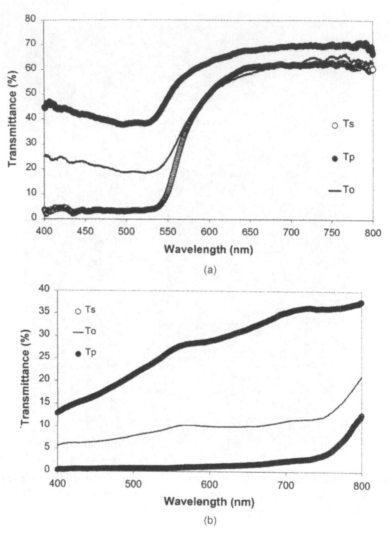

Figure 5.30. (a) Transmission spectra of the azo-dye AD-1 before and after exposure by a polarized light. T_p and T_s represent, respectively, the transmission spectrum perpendicular and parallel to the dye absorption axis, whereas T_o is the transmission spectrum before UV irradiation. (b) Transmission spectra of the azo-dye AD-1 after a special treatment with iodine or other vapours [105].

etc.) using a photo-mask. A latent image is invisible under normal conditions (figure 31(a)), but the image appears when viewed between crossed polarizers (figure 31(b)) or by using a single polarizer in the reflective case. Upon rotation or tilting the secure devices, the contrast of the image can be changed from positive to negative or one image can be changed to the

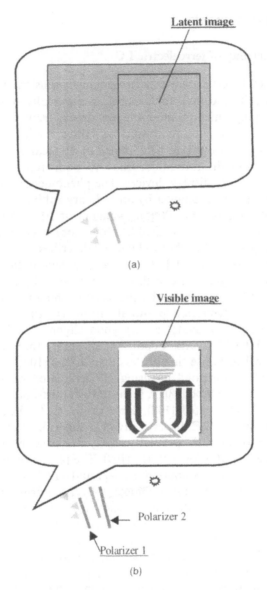

Figure 5.31. Latent (a) and visible colour (b) image, produced in photo-patterning photo-sensitive layer, such as PVCN or azo-dye material [93].

other. A colour shift and stereo-optical effects are possible. The optical security elements cannot be photocopied, but can be machine readable and personalized. The photo-patterning technology of photo-sensitive layers, such as PVCN, provides a high resolution of the security image [93].

5.5 New developments

5.5.1 Photo-aligning of ferroelectric LC

Ferroelectric LC (FLC) display cells are extremely sensitive to dust particles that induce nucleation centres for dislocations especially at a very small cell gap (less than 2 μm) which provides the best optical transmission of the cells [106]. Moreover, buffing non-uniformity very often results in the appearance of large domains with opposite FLC directors (typical dimension is about 0.5–3 cm). The domain formation leads to poor reproducibility in the manufacturing of FLC display devices. The phenomenon of the bistability degradation in FLC cells aligned by an ordinary rubbing technique is well known [28, 29]. This is why the photo-alignment FLC technology, which enables the avoidance of mechanical brushing, looks very promising.

The first paper on FLC photo-alignment describes the polyvinyl-alcohol layer doped with azo-dye [107]. The bistability degradation was observed even in the photo-aligned FLC cell, but it can be easily restored unlike in the case of an FLC cell oriented by the usual rubbing technique. A hybrid of linearly photo-polymerized photo-alignment (LPP) with liquid crystal polymer (LCP) layers provides a very good alignment of deformed helix ferroelectric (DHF) FLC with a contrast ratio more than 200:1 as well as the electro-optical response time of less than 200 μs [106]. Defect-free FLC displays were also fabricated by using UV irradiated polyimide films by the double exposure method [63]. A high contrast ratio and a perfect bistable switching was demonstrated [108]. Kobayashi *et al* fabricated a half-V-shaped photo-aligned FLC display, having a superior electro-optic performance in comparison with an FLC display prepared by rubbing with a high switching angle at a low voltage [109]. The photo-alignment proves to be useful also for polymer-stabilized V-shape and half-V-shaped ferroelectric displays due to a low operational voltage [110]. The normal sufficiently long UV illumination of polyimide film was shown to promote defect-free alignment of a surface-stabilized FLC layer due to a possible generation of a low pretilt angle of FLC on the substrate [111].

A remarkable property of azo-dye aligning layers is the pure reorientation of the molecular absorption oscillators perpendicular to the UV light polarization, which is not practically accompanied with photochemical transformations, as we stated above [12, 25]. This property gives a good chance of providing a high photo-alignment quality of FLC using the azo-dye SD-1 layer (figure 5.4). In experiment only one ITO surface of all prepared FLC cells was covered with an SD-1 layer, but another ITO surface was simply washed in *N,N*-dimethylformamide (DMF) [112]. The boundary condition asymmetry has been provided in order to avoid a competition in aligning action of solid surfaces of FLC cells. The photo-aligned FLC cell appeared to be better than that prepared by buffing, if the UV irradiation

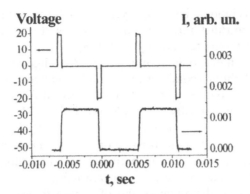

Figure 5.32. Perfect bistable switching of a photo-aligned FLC [112]. The bottom curve is the shape of the electro-optical response of the photoaligned FLC cell; the top curve is a driving voltage waveform.

time of the azo-dye SD-1 layer is high enough. The bistability switching was perfect and steady (figure 5.32); bistability degradation was not observed. Moreover, the multiplexing ability, which is necessary for a passively addressed FLC display cell operation, was also observed, due to existence of a certain threshold of FLC switching, $V_{th} \approx 0.6\,V$. The FLC response time of $\tau \approx 50\,\mu s$ at the voltage pulse amplitude of $U = 6\,V$ was demonstrated [112]. The new photo-aligning technology can be a decisive factor of the promotion of new FLC devices to the display market.

5.5.2 Photo-aligning of vertical aligned nematic (VAN) mode

VAN-LCD became most popular because of the high contrast and wide viewing angle [29]. VAN-LCD can be aligned even with a slantwise unpolarized light that makes this technique very promising for mass production applications [113–117]. At first the procedure of obtaining the VAN mode by a slantwise nonpolarized UV light using azobenzene side-chain polymer with a subsequent annealing was suggested by Ichimura *et al* [113]. The order parameter increases after the annealing procedure; however, it remains very small [113]. The same irradiation method by obliquely incident nonpolarized light allowed us to obtain slightly pretilted homeotropic alignment using polyimide films [114].

The photodegradation of commercially available polyimide aligning materials [115, 116] or crosslinking of photo-polymers [117, 118] during the exposure of obliquely incident unpolarized light is believed to be the main process responsible for the alignment in this case. Unfortunately the former process results in a decrease of the voltage holding ratio (VHR), which is a very important parameter especially for TFT-LCD fabrication. Moreover, notwithstanding great effort, the VAN-LCDs prepared by the photo-aligning technique have not yet reached the appropriate quality (response time, contrast ratio) in comparison with conventional LCDs, prepared by the rubbing technology [113–118].

The application of modern commercial VAN aligning agents with high resistivity to UV light in a combination with photo-aligned azo-dye materials [12, 25] can help to overcome the above-mentioned drawbacks. In experiment [119] the commercially available polyimide (PI) for homeotropic alignment was used in combination with azo-dye SD-1 [12, 25] for varying the pretilt angle from the homeotropic direction. The composition of 1% of SD-1 in solution with PI was prepared and the photo-aligning films were illuminated by a slantwise nonpolarized light.

For comparison, the aligning film of a pure PI was prepared by the rubbing technique to align the LC molecules in a homeotropic state with some pretilt angle. The measured value of the pretilt angle was about 1.4°, which is higher than in photo-aligned VAN LC cell (0.53°). However, the response time of the photo-aligned and conventional VAN LCD was almost the same, $\tau_{on} + \tau_{off} = 7.8\,ms + 9\,ms$ and $\tau_{on} + \tau_{off} = 8.1\,ms + 8.9\,ms$ respectively. Probably this means that SD-1 as a dopant in PI has increased further the azimuthal anchoring strength in the photo-aligned VAN-LCD cells. Moreover this increase takes place not only by the photo-degradation of PI in the UV light, but the pure reorientation of both SD-1 and PI molecules in such a manner that their absorption oscillators become perpendicular to the polarization of the activating UV light. The high values of VHR for the photo-aligned VAN-LCD using a SD-1/PI composition of 94–96% (close to the conventional rubbed VAN-LCD with VHR = 88%) testify to the point. The measured value of the contrast ratio between 'off' and 'on' states in all the cases exceeds 1000:1 in monochromatic light ($\lambda = 632.8\,nm$).

5.5.3 Photo-aligning of discotic LCs

As was mentioned above, the lyotropic LCs, which are columnar stacks of discotic molecules, can be photo-aligned, thus forming a thin polarizing layer (figures 5.26 and 5.27) [103, 104]. However, the thermotropic discotic LC can be also ordered by a photo-alignment method [120, 121]. A thin film of 25 nm thickness was prepared onto a fused silica surface from poly[4-(4-cyanophenylazo)phenyl methacrylate] polymer with side-chain p-cyanoazobenzene photosensitive groups (figure 5.33) [120]. The film was exposed to an obliquely incident non-polarized (incidence angle $\theta_i = 45°$) or normally incident polarized light ($\lambda = 436\,nm$, $E_{exp} = 3\,J/cm^2$) to provide oblique or normal alignment of disk-like molecules, respectively (figure 5.33). The order parameter and, consequently, the alignment quality increases by the subsequent annealing treatment of the azo-dye film [120]. Similar results of discotic alignment by a poly(vinylcinnamate) (PVCN) film, crosslinked

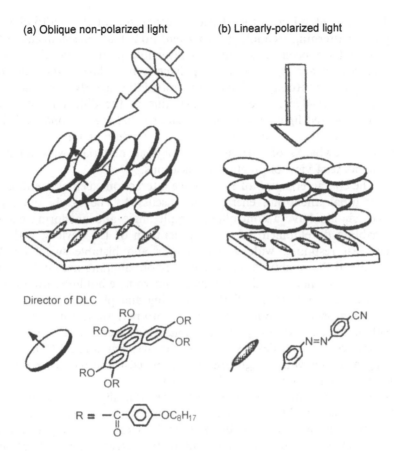

(a) Oblique non-polarized light   (b) Linearly-polarized light

Director of DLC

Figure 5.33. Tilted (a) and homeotropic (b) alignment of discotic LC (DLC) molecules (disks) onto a thin film of azo-benzene units (rods) exposed to irradiation obliquely with nonpolarized light and normally incident polarized light, respectively [121].

during the UV exposure, were also reported [121]. The obliquely irradiated nonpolarized light ($\lambda = 313\,\text{nm}$, $\theta_i = 45°$) was used for the purpose as in a previous case with exposure energy $E_{\exp} = 10\,\text{J/cm}^2$. The photo-alignment of thermotropic discotic LC (DLC) will help in the design of new optical retarders for TN-LCDs with wide viewing angles [29].

5.6   Conclusions

In this review we have briefly considered the novel photo-aligning and photo-patterning technology physics and their applications in LC devices. The review has described the following items.

1. An introduction to the physical mechanisms of the photo-aligning and photo-patterning technology was given. Four main mechanisms of the phenomena have been considered: (i) reversible photochemical *cis–trans* isomerization in azo-dye containing polymers, monolayers and pure dye films; (ii) pure photophysical reorientation of the azo-dye chromophore molecules or azo-dye molecular solvates due to the diffusion under the action of polarized light; (iii) topochemical crosslinking in cinnamoyl side-chain polymers; (iv) photodegradation in polyimide materials. The advantages and drawbacks of various photo-aligning materials were analysed from the point of view of practical applications.

2. The characterization of LC–surface interaction, such as pretilt angle and azimuthal anchoring energy, was discussed. The newly developed materials should have a controllable pretilt angle and anchoring energy, thus enabling development of a new generation of LC devices, with low voltage, fast response, wide viewing angles and bistable switching. We have analysed the quality development of photo-aligning materials, taking into account the thermo and UV stability and voltage holding ratio (VHR).

3. Various applications of photo-aligning and photo-patterning technology, such as multi-domain C cells, photo-patterning phase retarders and colour filters, and new patterned polarizing layers, were summarized. The examples of novel devices, such as wide-viewing-angle LCDs, colour birefringent LCDs with a high brightness, low-cost internal polarizers and a document security device, have been provided.

4. Photo-aligning of ferroelectric LC, vertical aligned nematic (VAN) modes and discotic LC has been described. The fast high-contrast low-voltage bistable switching in ferroelectric LCs can be a key factor for the new applications of these materials in the screens of mobile phones, e-books and microdisplays. The high-contrast photo-aligned wide-viewing-angle VAN mode can be successfully used for the large size TFT-LCD panels, used in LCD-TV and LCD monitors. The recent success in this field is encouraging.

Acknowledgments

We wish to thank the Hong Kong Research Grant Council for financial support under grant HKUST6004/01E.

References

[1] Kvasnikov E D, Kozenkov V M and Barachevskii V A 1977 *Dokl. Akad. Nauk SSSR* **237** 633 (in Russian)
[2] Ichimura K, Suzuki Y, Seki T, Hosoki A and Aoki K 1988 *Langmuir* **4** 1214

[3] Gibbons W M, Shannon P J, Sun S T and Swetlin B J 1991 *Nature* **351** 49

[4] Schadt M, Schmitt K, Kozenkov V and Chigrinov V 1992 *Jpn. J. Appl. Phys.* **31** 2155

[5] Dyadyusha A G, Marusii T Ya, Reznikov Yu A, Khizhnyak A I and Reshetnyak V Yu 1992 *JETP Lett.* **56** 17

[6] Hasegawa M and Taira Y 1995 *J. Photopolym. Sci. Technol.* **8** 241

[7] Blinov L M 1996 *J. Nonlinear Opt. Phys. Mat.* **5** 165

[8] O'Neill M and Kelly S M 2000 *J. Phys. D* **33** R67

[9] Schadt M 2001 *Mol. Cryst. Liq. Cryst.* **364** 151

[10] Kim J, Kim S, Park K and Kim T 2001 *SID '01 Digest* 806

[11] Pedersen T G, Ramanujam P S, Johansen P M and Hvilsted S 1998 *J. Opt. Soc. Amer.* **15** 2721

[12] Chigrinov V, Prudnikova E, Kozenkiv V, Ling Z, Kwok H, Akiyama H, Kawara T, Takada H and Takatsu H 2002 *SID '02 Digest* 1106

[13] Ichimura K, Hayashi Y, Akiyama H, Ikeda T and Ishizuki N 1993 *Appl. Phys. Lett.* **63** 449

[14] Aoki, K, Kawanishi Y, Seki T, Sakuragi M, Tamaki T and Ichimura K 1995 *Liq. Cryst.* **19** 119

[15] Aoki K, Seki T, Suzuki Y, Tamaki T, Hosoki A and Ichimura K 1992 *Langmuir* **8** 1007

[16] Knobloch H, Orendi H, Buchel M, Seki T, Ito Sh and Knoll W 1994 *J. Appl. Phys.* **76** 8212

[17] Liao Y, Hsu Ch and Wu S T 2000 *Jpn. J. Appl. Phys.* **39** L-90

[18] Ichimura K, Tomita H and Kudo K 1996 *Liq. Cryst.* **20** 161

[19] Shibaev V P, Kostromin S A and Ivanov S A 1996 in *Polymers as Electrooptical and Photooptical Active Media* (Berlin: Springer) pp 37–110

[20] Jones C and Day S 1991 *Nature* **351** 15

[21] Rosenhauer, Fischer Th, Czapla S, Stumpe J, Vinuales A, Pinol M and Serrano J L 2001 *Mol. Cryst. Liq. Cryst.* **364** 295

[22] Sun S T, Gibbons W M and Shannon P J 1992 *Liq. Cryst.* **12** 869

[23] Shannon P J, Gibbons W M and Sun S T 1994 *Letters to Nature* **368** 532

[24] Gibbons W M, Kosa T, Palffy-Muhoray P, Shannon P J and Sun S T 1995 *Letters to Nature* **377** 43

[25] Chigrinov V G, Kwok H S, Yip W C, Kozenkov V M, Prudnikova E K, Tang B Z and Salhi F 2001 *Proc. SPIE* **4463** 117

[26] Kozenkov V M, Yudin S G, Katyshev E G, Palto S P, Lazareva V T and Barachevskiy V A 1986 *Pisma J. Theor. Fiz.* **12** 1267 (in Russian)

[27] Barnik M I, Kozenkov V M, Shtykov N M, Palto S P and Yudin S G 1989 *J. Mol. Electr.* **5** 53

[28] Blinov L M and Chigrinov V G 1994 *Electrooptic Effects in Liquid Crystal Materials* (Berlin: Springer)

[29] Chigrinov V G 1999 *Liquid Crystal Devices: Physics and Applications* (London: Artech House).

[30] Kwok H S and Chigrinov V G 2002 *SPIE Conference 4799* 4799-28 July 2002

[31] Pedersen T G and Johansen P M 1997 *Phys. Rev. Lett.* **79** 2470

[32] Egerton P G, Pitts E and Resier A 1981 *Macromolecules* **14** 95

[33] Obi M, Morino S and Ichimura K 1999 *Jap. J. Appl. Phys.* **38** L145

[34] Kawatsuki N, Goto K and Yamamoto T 2001 *Liq. Cryst.* **28** 1171

[35] Schadt M, Seiberle H and Schuster A 1996 *Nature* **351** 212

[36] Kawatsuki N, Yamamoto T and Ono H 1999 *Appl. Phys. Lett.* **7** 935
[37] Yaroshchuk O, Cada L G, Sonpatki M and Chien L-C 2001 *Appl. Phys. Lett.* **79** 30
[38] Yaroshchuk O, Pelzl G, Pirwitz G, Reznikov Yu, Zaschke H, Kim J-H and Kwon S B 1997 *Jpn. J. Appl. Phys.* **36** 5693
[39] Kimura M, Nakata Sh, Makita Y, Matsuki Y, Kumano A, Takeuchi Y and Yokoyama H 2001 *Jpn. J. Appl. Phys.* **40** L352
[40] Perny S, Barny L E, Delaire J, Buffeteau T and Sourisseau C 2002 *Liq. Cryst.* **27** 341
[41] West J L, Wang X, Li Y and Kelly J R 1995 *SID '95 Digest* 703
[42] Nishikawa M, Taheri B and West J L 1998 *Appl. Phys. Lett.* **72** 2403
[43] Sakamoto K, Usami K, Watanabe M, Arafune R and Ushioda S 1998 *Appl. Phys. Lett.* **72** 1832
[44] Gong S, Kanicki J, Ma L and Zhong J 1999 *Jpn. J. Appl. Phys.* **38** 5996
[45] Sung S-J, Kim H-T, Lee J-W and Park J-K 2001 *Synth. Met.* **117** 277
[46] Kozenkov V M and Barachevskii V A 1987 *Organic Photoanisotropic Materials and Their Application* ed V A Barachevsky (Leningrad: Nauka) p 89 (in Russian)
[47] Yang K H, Tajima K, Takenaka A and Takano H 1996 *Jpn. J. Appl. Phys.* **35** L561
[48] Wang Y, Xu C, Kanazawa A, Shiono T, Ikeda T, Matsuki Y and Takeuchi Y 2001 *Liq. Cryst.* **28** 473
[49] Kim J-H, Acharya B R, Agra D M and Kumar S 2001 *Jpn. J. Appl. Phys.* **40** 2381
[50] Wang Y, Natsui T, Makita Y, Kumano A and Takeuchi Y 2000 *IDW '00 Digest* 57
[51] Lee W-C, Hsu C-S and Wu S-T 2000 *Jpn. J. Appl. Phys.* **39** L471
[52] Kim H-T, Lee J-W, Sung S-J and Park J-K 2000 *Liq. Cryst.* **27** 1343
[53] Jain S C, Rajesch K, Samanta S B and Narikar A V 1995 *Appl. Phys. Lett.* **67** 1527
[54] Bryan-Brown G P and Sage I C 1996 *Liq. Cryst.* **20** 825
[55] Choi Y-J, Kim K-J, Choi Y-S and Kwon S-B 1997 *SID '97 Digest* L-25
[56] Dyadyusha A, Khizhnyak A, Marusii T, Reznokiv Yu, Yaroschuk O, Reshetnyak V, Park W, Kwon S, Shin H and Kang D 1995 *Mol. Cryst. Liq. Cryst.* **263** 399
[57] Schadt M, Seiberle H, Schuster A and Kelly S M 1995 *Jpn. J. Appl. Phys.* **34** 3240
[58] Schadt M, Seiberle H, Schuster A and Kelly S M 1995 *Jpn. J. Appl. Phys.* **34** L764
[59] Seiberle H, Muller O, Marck G and Schadt M 2002 *Journal of the SID* **10/1** 31
[60] Gibbons W M, McGinnis B P, Shannon P J, Sun S T and Zheng H 1999 *IDW '99 Digest* p 81
[61] Chigrinov V, Prudnikova E, Kozenkov V, Kwok H, Akiyama H, Kawara T, Takada H and Takatsu H 2002 *Liq. Cryst.* accepted for publication
[62] Dyadyusha A, Khizhnyak A, Marusii T, Reshetnyak V, Reznikov Yu and Park W-S 1995 *Jpn. J. Appl. Phys.* **34** L1000
[63] Iimura Y, Saitoh T, Kobayashi S and Hashimoto T 1995 *J. Photopolym. Sci. Tech.* **8** 258
[64] Yamamoto T, Hasegawa M and Hatoh H 1996 *SID '96 Digest* 642
[65] Seo D-S, Hwang L-Y and Kobayashi S 1997 *Liq. Cryst.* **23** 923
[66] Park B, Han K-J, Jung Y, Choi H-H, Hwang H-K, Lee S, Jang S-H and Takezoe H 1999 *J. Appl. Phys.* **86** 1854
[67] Kim M W, Rastegar A, Olenik I and Rasing T 2001 *J. Appl. Phys.* **90** 3332
[68] Lazarev V V, Barberi R, Iovane M, Papalino L and Blinov L M 2002 *Liq. Cryst.* **29** 273
[69] Shenoy D, Beresnev L, Holt D and Shashidhar R 2002 *Appl. Phys. Lett.* **80** 1538
[70] Vorflusev V P, Kitzerow H S and Chigrinov V G 1995 *Jpn. J. Appl. Phys.* **34** L1137

[71] Vorflusev V P, Kitzerow H S and Chigrinov V G 1997 *Appl. Phys.* **A64** 615
[72] Chen J, Bos P J, Bryant D R, Johnson D L, Jamal S H and Kelly J R 1995 *Appl. Phys. Lett.* **67** 1990
[73] Aerle N A 1995 *Jpn. J. Appl. Phys.* **34** L1472
[74] Cheng S-D and Sun Z-M 1995 *Liq. Cryst.* **19** 321
[75] Schadt M, Seiberle H and Schuster A 1996 *Letters to Nature* **381** 212
[76] Konovalov V A, Minko A A, Muravski A A, Timofeev S N and Yakovenko S E 1999 *J. of the SID.* **7/3** 213
[77] Park J-H, Yoon T-Y, Lee W-J and Lee S-D 2001 *IDW '01 Digest* 221
[78] Park J-H, Choi Y, Yoon T-Y, Yu C-J and Lee S-D 2002 *ASID '02 Digest* 367
[79] Wang X, Subacius D, Lavrentovich O, West J and Reznikov Y 1996 *SID '96 Digest* 654
[80] Kim J H, Yoon K H, Wu J W, Choi Y J, Nam M S, Kim J and Kwon S B 1996 *SID '96 Digest* 646
[81] Soh H S, Wu J W, Nam M S, Choi Y J, Kim J, Kim K J, Kim J H and Kwon S B 1996 *Euro Display '96 Digest* 579
[82] Schadt M and Seiberle H 1997 *SID '97 Digest* 397
[83] Lee J S, Han K Y, Chae B H, Park G B and Park W S 1998 *Asia Display '98 Digest* 781
[84] Seiberle H and Schadt M 1998 *Asia Display '98 Digest* 193
[85] Seiberle H, Schmitt K and Schadt M 1999 *Euro Display '99 Digest* 121
[86] Hoffmann E, Klausmann H, Ginter E, Knoll P M, Seiberle H and Schadt M 1998 *SID '98 Digest* 734
[87] Nishikawa M, Taheri B, West J and Reznikov Y 1998 *Jpn. J. Appl. Phys.* **37** L1393
[88] Kvasnikov E D, Kozenkov V M and Barachevskii V A 1978 *Nonsilver and Unusual Media for Holography* 96 (in Russian)
[89] Shulev Yu V, Kozenkov V M and Barachevskii V A 1983 *Photochemical Processes of Hologram Recording* ed. V A Barachevskii, Leningrad, p 47 (in Russian)
[90] Shulev Yu V, Kozenkov V M, Barachevskii V A *et al* 1986 *Integrated Optics: Physical Bases, Application* (Novosibirsk: Nauka) p 83 (in Russian)
[91] Kozenkov V M, Kvasnikov E D, Barachevskii V A *et al* 1980 *Pisma J. Tehn. Fiz.* **6** 105 (in Russian)
[92] Kozenkov V M, Odinokov S B, Petrushko I V *et al* 1982 *Optical and opto-electronic methods of image processing* (Leningrad: Nauka) p 82 (in Russian)
[93] Kozenkov V M, Chigrinov V G and Kwok H S 2002 accepted for publication in *Mol. Cryst. Liq. Cryst.*
[94] Yakovlev D A, Simonenko G V, Kozenkov V M, Chigrinov V G and Schadt M 1993 *Euro Display '93 Digest* 17
[95] Schadt M, Seiberle H and Moia F 1999 *IDW '99 Digest* 1013
[96] Chen J, Chang K C, DelPico J, Seiberle H and Schadt M 1999 *SID '99 Digest* 98
[97] http://www.rolic.com
[98] DelPico J, Chang K-C and Sharp G P 2000 *Information Display* **7/00** 28
[99] Cobb C 2000 *Seminar presentation of OPTIVA Inc*, February 2000, Hong Kong
[100] Bobrov Y, Cobb C, Lazarev P, Bos P, Bryant D and Wondely H 2000 *SID '00 Digest* 1102
[101] Sergan T, Schneider T, Kelly J and Lavrentovich O D 2000 *Liq. Cryst.* **27** 567
[102] Khan I G, Belyaev S V, Malimonenko N V, Kukushkina M L, Shishkina E Yu, Masanova N N and Vorozhtsov G N 2002 *Euro Display '02 Digest* 573

[103] Kozenkov V M, Yip W C, Tang S T, Chigrinov V G and Kwok H S 2000 *SID '00 Digest* 1099

[104] Yip W C, Kwok H S, Kozenkov V M and Chigrinov V G 2001 *Displays* **22** 27

[105] Kozenkov V M, Yip W C, Chigrinov V, Prudnikova E and Kwok H S 2001 *Photo-induced dichroic polarizers and fabrication methods thereof*, HKUST patent, pending in March, 2001

[106] Funfschilling J, Stalder M and Schadt M 1999 *SID '99 Digest* 308

[107] Vorflusev V, Kozenkov V and Chigrinov V 1995 *Mol. Cryst. Liq. Cryst.* **263** 577

[108] Kurihara R, Furue H, Takahashi T and Kobayashi S 1999 *IDW '99 Digest* 93

[109] Murakami Y, Kawamoto S, Xu J and Kobayashi S 2001 *IDW '01 Digest* 93

[110] Murakami Y, Xu J, Kobayashi S, Endo H and Fukuro H 2002 *SID '02 Digest* 496

[111] Kang W-S, Kim H-W and Kim J-D 2001 *Liq. Cryst.* **28** 1715

[112] Pozhidaev E P, Chigrinov V G, Huang D D and Kwok H S 2002 *Euro Display '02 Digest* 137

[113] Furumi S, Nakagawa M, Morino S, Ichimura K and Ogasawara H 1999 *Appl. Phys. Lett.* 7 2438

[114] Ushinohama H, Yoshikawa H, Furue H, Takahashi T and Kobayashi S 1999 *IDW '99 Digest* 285

[115] Seo D-S, Park D-S and Jeon H-J 2000 *Liq. Cryst.* **27** 1189

[116] Hwang J-H and Seo D-S 2001 *Jpn. J. Appl. Phys.* **40** 4160

[117] Hwang J-Y and Seo D-S 2001 *Liq. Cryst.* **28** 1065

[118] Hwang J-H, Seo D-S, Kim J-Y and Kim T-H 2002 *Jpn. J. Appl. Phys.* **41** L58

[119] Konovalov V A, Chigrinov V G and Kwok H S 2002 *Euro Display '02 Digest* 529

[120] Furumi S, Ichimura K, Sata H and Nishiura Y 2000 *Appl. Phys. Lett.* **17** 2689

[121] Ichimura K, Furumi S, Morino S, Kidowaki M, Nakagawa M, Ogawa M and Nishiura Y 2000 *Adv. Mater.* **12** 950

Chapter 6

Industrial and engineering aspects of LC applications

Tomio Sonehara

6.1 Practical spatial modulation

Liquid crystal–spatial light modulators (LC-SLMs) are essentially not capable of high-speed switching compared with semiconductor-based spatial light modulators (SLMs) or optical crystals. The LC-SLMs, however, are one of the most valuable devices from the point of view of easy handling, good reliability and large pixel capacity, based on liquid crystal display (LCD) technologies.

For practical applications, such as displays and light modulations, the methods of spatial addressing have been sought for a long time. These methods can be used for scanning technologies like an electron beam for cathode ray tubes (CRTs) or laser beams for laser displays, which are a basic concept of image formation. Current addressing technologies, passive and active matrix, are sophisticated technologies making full use of liquid crystal optical modulation characteristics. They have, however, been developed only for displays. Consequently we should well understand their driving features when using them as a spatial light modulator. In the first half of the chapter, the spatial addressing technologies in practical use and historical views are described.

For display applications, only twisted nematic (TN) and related modes have been substantially used for the past 30 years. The function utilized for these applications should be a part of the light-modulating ability of liquid crystal.

In the latter half of the chapter, other applications are described, bringing out other optical modulating functions of liquid crystal, the first of which is the amplitude modulation function, which is used as light valves employed in most liquid crystal (LC) projectors. Another function of LCs is phase modulation. Liquid crystal has relatively large birefringence compared with inorganic materials. If high-resolution two-dimensional phase modulators using the birefringence are developed, then holographic image reconstruction and/or electrically controlled holographic optical elements (HOEs) can be easily realized. Finally, scattering or deflection modulation, for instance polymer dispersed liquid crystal (PDLC)—a kind of LC composite system—and its characteristics are described.

6.2   Spatial addressing technologies

Since the era of ancient wall painting, optical changing has been making an image. Chalk drawing uses a scattering variation on a slate, and painting is based on the absorption of pigments on a canvas. All of them are static spatial modulation of light by using two-dimensional addressing by hand. They are categorized into vector-scanning and raster-scanning, which correspond to drawing at random or regularly. In the history of display devices, radars and graphic displays utilized vector scanning for electron-beam accumulation on a CRT. Nowadays, because most devices should handle two-dimensional graphical images, almost all display devices may be categorized as raster-scanning types. Accordingly, current addressing technologies for LCDs should also be raster-scanning.

Some use the word 'addressing' to mean scanning, and others use it to mean driving. In this chapter, addressing is defined with two functions, which are assignment and distribution of data.

6.2.1   Passive matrix addressing

Passive matrix addressing had been developed on the way from segment-type displays to graphic displays. It is based on a multiplexing concept, which developed electro-optical modulators, like $LiNbO_3$ crystals, to reduce the driving circuits by utilizing its steep field-response characteristics.

Passive matrix addressing is commonly interpreted as an x–y matrix or a simple matrix, which is the structure of two substrates facing each other with stripe-shaped electrodes, with LC material placed between them (figure 6.1). Super-twisted nematic (STN) and related modes have also utilized passive addressing. The modes are usually used in relatively small-sized display devices, and have some drawbacks as mentioned below.

If LC media have intrinsic memory and fast response, passive addressing is the simplest and the most versatile method. The intrinsic memory enables infinite addressing lines, because any unselected lines do not affect a selected line. While the memory conveys completely independent addressing of each other, fast response enables more lines in a limited frame time. As a matter of fact, there is no such ideal LC medium. Among LC materials and modes, surface-stabilized ferroelectric liquid crystal (SSFLC) would be close to the ideal. However, the reason why most LC displays

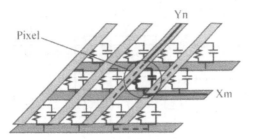

Figure 6.1. Passive addressing.

have not employed the FLC modes are the difficulties of fabrication and less grey-scale capability. Some devices for scientific research are commercially available [1].

6.2.1.1 Multiplexing addressing

Passive addressing using multiplexing is complicated compared with active matrix addressing, because a pixel has no storage data function. Passive addressing, consequently, needs data accumulation in its driving circuits. Another reason why it is complicated is that LC responds nonlinearly in accordance with the applied effective voltage that must be calculated over a frame. Each pixel is connected to a selected line and a data line, and is also connected to all selected lines and data lines through neighbouring pixels. This equivalent circuit, which consists of many capacitor, C, and resistor, R, elements, looks three-dimensional (figure 6.2) [2]. The mutual relations of all pixels cause cross talk, which degrades display quality with blurred and faded images, and slow response time. The cross-talk results in limiting of the number of addressing lines. Consequently, passive addressing puts a limit on the number of lines. However, it can easily obtain a pixel with a high aperture ratio.

Otherwise, passive addressing is the most sophisticated technology among other technologies. Looking at figure 6.2, the most simple method is as follows. Select a line (V_{select} applied) and data (0 representing V_{data}, 1 representing $-V_{data}$) are written to one line at a time. This is called the 'line at a time method'. According to the LC nonlinear optical response, the effective voltage (V_{eff}) is calculated as the waveform applied between x and y electrodes and can select the ON or OFF states.

If instead several lines are selected at a time, the calculation of data must be complicated and data shape would be jagged. This is called multi-line addressing (MLA) [3]. When all lines are selected at a time, all line data should be calculated through a huge multiply and accumulation with ortho-gonal functions. This is called active addressing [4]. Please note, this is not the active matrix addressing described in the next section.

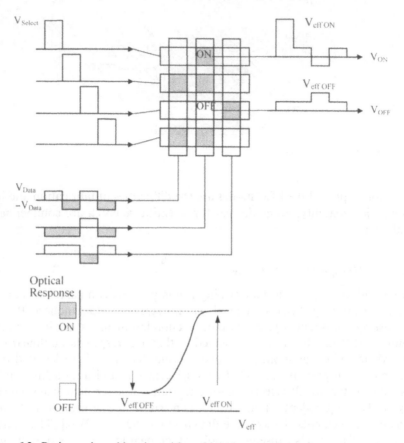

Figure 6.2. Basic passive addressing with multiplexing waveform.

6.2.2 Active matrix addressing

Active matrix addressing is widely used in display devices, which are integrated active electrical elements such as thin film transistor (TFT), liquid crystal on silicon (LCOS), thin film diode (TFD), metal insulator metal (MIM), etc. Almost all have exactly similar operation; that is, the storage of electric charge in each pixel during a frame. Each pixel can be addressed independently, so that the limit of the addressing line may be solved. For the application of SLMs, the addressing is the most important means of realizing various optical functions. Table 6.1 shows typical active matrix LCDs which are applicable to SLM applications and are commercially available.

6.2.2.1 TFT active matrix

TFTs are the most frequently used devices for PC displays, liquid crystal televisions (LCTVs), mobile equipment and LC projectors. Figure 6.3 depicts

Table 6.1. Active matrix LCDs applicable to SLMs.

	Substrate	Modulation mode	Pixel pitch (μm)	Resolution	Diagonal size (inches)	Response Time
IBM*	x-Si	TN-ECB (reflective)	17	2048 × 2048	1.9	Video rate
JVC-Hughes	x-Si	⊥-ECB (reflective)	13	1365 × 1024	0.9	Video rate
Hitachi	x-Si	TN-ECB (reflective)	8.1	1920 × 1080	0.7	Video rate
Philips FDS	x-Si	TN-ECB (reflective)	18	1024 × 768	0.97	Video rate
Three-Five	x-Si	TN-ECB (reflective)	9.5	1920 × 1200	0.85	Video rate
Displaytech	x-Si	Ferro. LC (reflective)	12	320 × 240	0.19	100 Hz
MIT*	x-Si	Ferro. LC (reflective)	10	800 × 600	0.4	70 μs
Kopin	x-Si film	TN (transmissive)	15	1280 × 1024	0.96	180 frames/sec
Seiko Epson	p-Si TFT	TN (transmissive)	14	1024 × 768	0.7	Video rate
Sony	p-Si TFT	TN (transmissive)	26	1024 × 768	1.3	Video rate
Sharp	L pSi TFT	TN (transmissive)	45, 32	1280 × 1024	2.6	Video rate
TI DMD	x-Si	Mirror (not LC)	17	1024 × 768	0.9	120 frames/sec

* According to development reports.

x-Si: single-crystal Si; p-Si TFT: poly-Si thin film transistor; L-pSi TFT: low-temperature poly-Si thin film transistor; TN-EBC: twisted nematic electrically-controlled birefringence; ⊥-EBC: perpendicularly-aligned electrically-controlled birefringence; PDLC: polymer dispersed liquid crystal; Ferro. LC: ferroelectric liquid crystal.

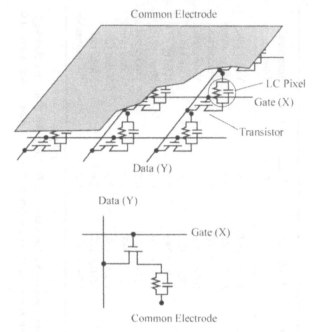

Figure 6.3. TFT active matrix.

a typical equivalent circuit of a TFT active matrix. The TFT's drain is connected to the LC pixel and the source to a data electrode. The gate and data lines form the x–y matrix. The TFTs are usually categorized as amorphous silicon (a-Si) TFT, poly-Si TFT, low temperature (below 600°C), and so on. They are differentiated by the TFT fabrication process, especially the deposition temperature of Si.

The mobility of Si directly decides TFT characteristics, such as switching speed and dark current. a-Si TFTs are not capable of integrating a built-in driving circuit on the glass substrate.

The process temperature is the most severe constraint for substrate material because economic transparent glass substrates are limited. Usually low-alkaline glass is used for an a-Si TFT and low-temperature poly-Si, and quartz for poly-Si.

For direct view LCDs, a-Si TFTs are utilized due to the requirements of low cost and wider display area. On the other hand, for SLM applications poly-Si TFT is likely to be used due to high-density pixel.

Figure 6.4 depicts the TFT-driven phase SLM developed by us. It incorporates poly-Si TFTs as switching elements, and integrated poly-Si TFT drivers on the same glass substrate. Figure 6.4 is a cross-sectional view of a pixel area. Figure 6.5 is a typical light-shielding pattern. TFT, even made of poly-Si, will leak photocurrent under ultra-violet (UV) light irradiation. To avoid the leakage, the light shield is usually employed on

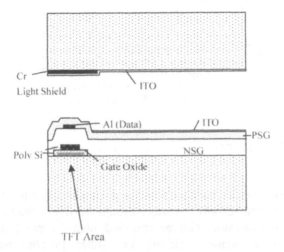

Figure 6.4. Typical TFT active matrix structure.

the counter substrate. The pattern affects aperture ratio and the diffraction characteristics. Usually, it is made of a metal or a pigment layer.

LC should be driven with an a.c. field to avoid degradation by a d.c. field. The alternation must affect spatial modulation characteristics in actual experiments. When we use the TFT device as an SLM, we should understand the difference of driving methods, which are alternating methods, and grey-level representation with digital or analogue.

6.2.2.2 LCOS and Si-based matrix

LCOS literally means 'on silicon'. It is a reflective modulator that employs reflective electrodes and a reflective LC mode. Its basic driving methods

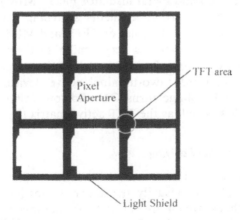

Figure 6.5. Light shield pattern.

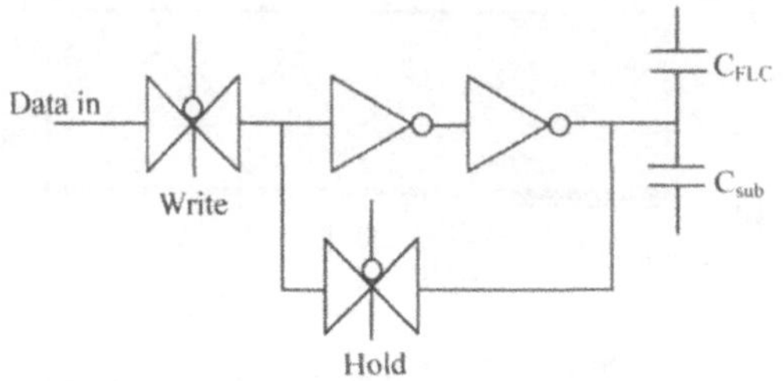

Figure 6.6. SRAM-type pixel.

are the same as the TFT's, by which charge memory at a pixel is carried out. On the other hand, LCOS has more design flexibility than a TFT. This is because more transistors can be installed below a pixel electrode. Thus another type of memory circuit, i.e. static random access memory (SRAM), may be easily built there. Some LCOS devices employ the SRAM function instead of the charge memory, which integrates flip-flop logic at each pixel. This brings more uniform spatial modulation compared with TFTs (figure 6.6). The pixel density of TFTs and LCOS is limited by the size of transistors, which depend on the state of art of the fabrication process.

Recently, to improve transistors themselves, some unique technologies have been developed. Among these are implant technologies where Si thin film peeled from a Si wafer or a TFT glass substrate is transplanted onto a separate transparent substrate. EPSON and Kopin have been developing this technique [5, 6].

6.2.2.3 Two-terminal active matrix

Thin film diode (TFD) and metal insulator metal (MIM) are two-terminal active elements, which work as elements switching between charge and discharge of the pixel. Here the roles of TFDs and MIMs is similar to that of TFTs. Further, their matrix is very similar to the x–y passive matrix structure. The cross portion of x–y electrodes behaves as a pixel, shown in figure 6.7. In other words, two-terminal active devices work like diode switches, where high voltage turns ON and low voltage OFF. They are applicable to SLMs as well as the TFT active matrix [7].

6.2.2.4 Alternating field driving

As mentioned above, LCs have to be driven with an a.c. field to avoid electrochemical degradation. Usually the active addressing utilizes three a.c. field-driving methods, which are called F (frame) inversion, H (horizontal line) inversion and dot inversion (figure 6.8). In the case of SLM applications,

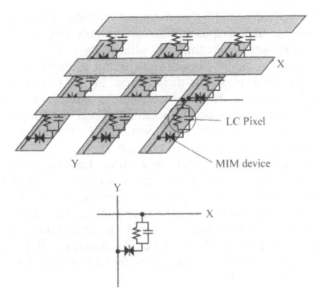

Figure 6.7. MIM active matrix.

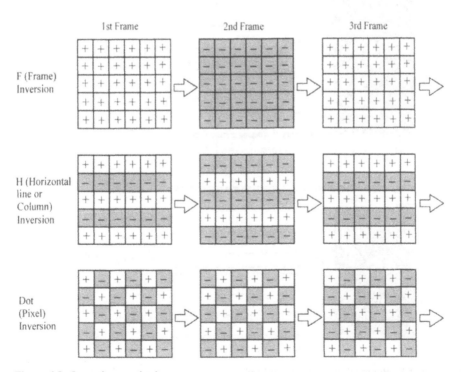

Figure 6.8. Inversion methods.

the inversion method would be more important than display applications. This is because the SLM requires more optical uniformity in the pixel area and different response time compared with displays. F inversion is more uniform in optical response. Response variation among frames, however, exists. Some LCOS displays use this method. Most LCDs employ H and dot inversion in practice. H and dot inversion suffer from a lateral field effect, i.e. the surrounding pixels' potential affects the longitudinal field uniformity in the pixel. Moreover, the potential of scan lines and data lines also affects the field uniformity. Accordingly as the size of pixel becomes tiny as the lateral field effect should be counted. Looking at fluctuation of the optical response, F inversion is sensitive for the invariance between + and − polarity. Recent LC display devices employ a black mask light shield layer, which prevents the leakage of light through the surrounding areas of each pixel, and the mask can make contrast ratio and extinction ratio higher. In other words, the mask can shield the optical non-uniformity in the pixel, which is caused by a lateral electric field between the gate and the pixel potential (figure 6.9).

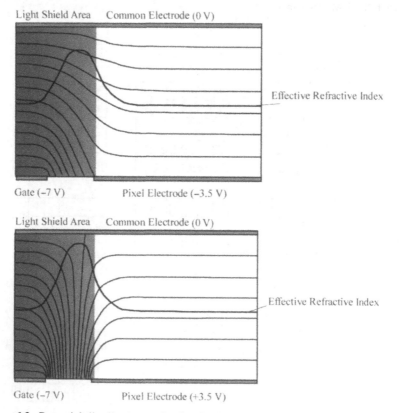

Figure 6.9. Potential distribution and refractive index gradient.

6.2.2.5 Charge memory and intrinsic memory

Most LCDs have been developed for display devices. Accordingly, they only take into account human eye response. This is the technical basis of CRT displays and video formats such as NTSC and PAL. LC display devices should have been developed in accordance with it. When we want to use LCDs as SLMs, we face some difficulties, such as non-uniformities in time and space.

As for dynamic response, there are two kinds, which are due to fast LC fluctuation and slow average response. On microscopic viewing, liquid crystal generally can respond up to a cut-off frequency, which is called the dielectric relaxation frequency. It is about 1–10 kHz in the case of standard nematic liquid crystals. For display applications, you cannot see the fluctuation because of your eye's response. In the case of passive addressing SLMs, a high-frequency driving waveform tends to cause fluctuation.

Another is due to whole response of the LC layer. In the case of active matrix, it depends on both memory method and LC response.

The charge stored at a pixel gradually decays during a frame time. The decay time is decided by the CR constant, where C is the sum of all capacitive elements connected to a pixel and R is the sum of series resistive elements. Since current TFT LCDs incorporate a large additional capacitor connected to a pixel, the optical fluctuation can be neglected. However, optical response caused by the response of the whole LC layer varies gradually over a few frames. It is called accumulated response effect. Thus you should comprehend the dynamic properties, considering what response you need.

Intrinsic LC memory, such as ferroelectric LC, shows stable modulation after addressing.

Thus memory method is crucial to obtain uniform modulation in the time dimension.

6.2.2.6 Longitudinal and lateral field effect

LC material has a large dielectric and refractive index anisotropy. This permits thin-film devices with longitudinal field effects to be made. In the case of a high-density pixel, the effect is strongly affected by the lateral field effect. In addition, using TN configuration, transmitted light is varied with viewing angle. To improve the uniformity of viewing contrast ratio from any viewpoint, lateral field effects have been developed. They employ comb electrodes called in-plane switching (IPS) [8] and fringe field switching (FFS), for example (figure 6.10) [9]. It is also preferable for LCD assembly that LC molecules move on a plane, because no electrode forms on the counter substrate. The comb electrode itself has a long history. Early electro-optical (EO) modulators using an EO crystal such as $LiNbO_3$

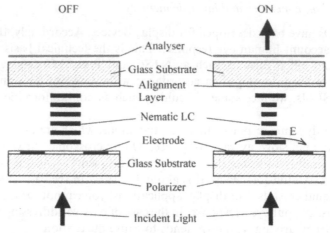

Figure 6.10. Lateral field mode (example: IPS).

employed the comb electrode because a relatively large thickness was required for light modulation. There seems to be a revival of the comb.

6.2.3 Optical addressing

6.2.3.1 Thermal optical addressing [10]

Optical addressing is a similar method to photography. In the early stages of LCD development, directly writing technology was developed. It is a method of laser drawing on an LC medium by irradiating thermal energy of addressing light. Smectic A (SmA) LC and cholesteric LC were used as the media, whose thermal phase change or thermal change in LC alignment are utilized. In principle, while it can realize complete memory and independent pixel addressing, it cannot realize moving modulation due to its slow response time. Because it has no limitation of pixel numbers, it was used as a large-size display or a rewritable photo-mask. Using the technology, the application of an electronic paper is expected soon (figure 6.11).

6.2.3.2 Photoconductor addressing

Because optical addressing is secondary, a primary addressing is required, such as CRT or LCD imaging devices. There is no discrete pixel structure in the optical addressing device, which should solve a particular issue of pixel separation. In other words, it is how to manage charge in the system. Most optical addressing devices utilize a dielectric mirror for separation between writing and reading light. Figure 6.12 is a structure using high resistive Si as the photoconductor. The mirror, photoconductor and LC

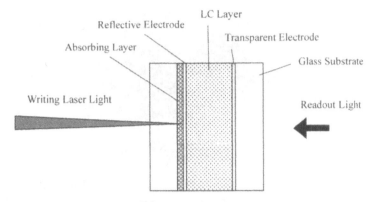

Figure 6.11. Thermal optical addressing.

material should be designed taking into consideration the electrical impedance distribution of each layer. If there is no dielectric mirror, you can attain almost the same resolution as photoconductor thickness.

Hughe developed the first optical addressing device for projector systems with high intensity. The light emitted from addressing a CRT stimulates a photoconductor, and it duplicates the light image in an LC layer in two dimensions. An initial model incorporated a 45° TN mode and CdS for the photoconductor [11]. The configuration of the device has been changed to an a-Si and ECB (electrically controlled birefringence) mode [12] currently, and most devices continue using nematic LC. These devices exhibit relatively high resolution, 10–50 μm, which is limited by a dielectric mirror placed between the LC and the photoconductor.

For projector applications, optical addressing has been replaced by liquid crystal on silicon (LCOS), because LCOS can obtain more than 1 M

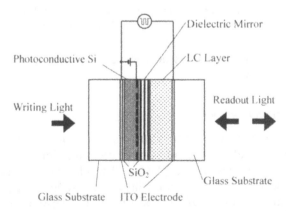

Figure 6.12. Optically addressed SLM (example: Si-based type).

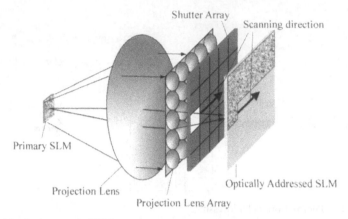

Figure 6.13. A giant-scale SLM.

pixels and a separation less than 20 μm. Instead of the projector, the photo addressing system is utilized for infrared (IR) imaging and optical data processing. For instance, InGaAs as a photo conductor has sensitivity in the IR. Hamamatsu has offered nematic LC photo-addressed SLMs for phase modulation. Other LC modes, SSFLC [13] and PDLC [14], have been used. Recently, the same kind of giant-scale system, with 10^{12} pixels, has been developed by DERA [15]. This is for real-time holographic reconstruction (figure 6.13).

6.2.4   Other addressing

6.2.4.1   Electron beam addressing

The concept of electron beam addressing is similar to a cathode ray tube (CRT). An accelerated electron beam collides with a thin glass plate as an anode, and the accumulated electrons form a secondary image on the glass. Finally the secondary image is transferred to an LC layer as potential distribution (figure 6.14) [16]. It is almost the same technology as Eidophor [17], which was widely used as a large-venue projector and based on oil membrane distortion.

6.2.4.2   Plasma addressing

The basic structure is similar to a plasma discharge display panel (PDP). However, plasma addressing does not use radiation. It uses only the threshold characteristics of the plasma. The plasma is placed between a thin glass substrate adjacent to an LC layer and a supporting glass substrate. The plasma image can be duplicated on the LC layer by controlling the potential of the thin glass substrate. This technology has been developed

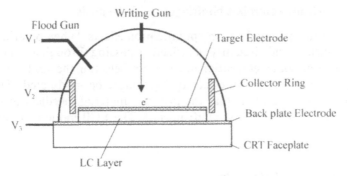

Figure 6.14. Electron beam addressing.

to fabricate a large-sized display panel because TFT has process limitations for a large-sized substrate. Consequently, it would be difficult to get high pixel density due to fabricating discharge space between substrates.

6.2.5 Future aspect

Here, I would like to look at future aspects and predict future developments. One of the prospects could be found in the light valve, in particular for projection display, because light valves require excellent quality, fast response, optical uniformity, high extinction ratio, rich pixel number and need digital driving. This direction is also preferable for SLM applications because any light modulation in general must require deep modulation depth, fast response and linearity. In the future, display technologies should lead the way for the SLM. LCOS and TFT in particular would expand their applicable area towards general optical modulations, such as a phase modulator or a directional modulator. We can find heralds of such new applications. Deducing the specifications according to TFT-LCD development speed, we may forecast a pixel pitch of less than 5 μm in the near future.

6.3 Amplitude modulation and applications

6.3.1 Twisted nematic (TN) mode

LC displays are two-dimensional amplitude modulation as light intensity. Usually they employ micro red, green and blue (RGB) colour filters for colour reproduction. The most common displaying mode is TN. Using relatively high refractive index difference, Δn, even a few-micron LC film exhibits large polarization circulation. In this sense, TN is an epoch-making invention.

6.3.2 Electrically controlled birefringence (ECB) mode

ECB is the principal mode in any polarization effects. When polarized light, including linearly and circularly polarized, is passing through an anisotropic optical medium, some retardation must occur. Selecting the kind of polarized light by an analyser, the intensity of light must be modulated. The most important feature of LC is that the precious material, needless to say, can easily obtain 0.2 more of the birefringence. The ECB mode is usually used for SLMs.

6.3.3 Guest–host (GH) mode

This mode has been used as a date indicator in compact cameras utilizing high extinction ratio with dense dye content. Dyestuff can depict a natural absorption image, like printing. In the GH system, the LC works as an initiator of dye molecule alignment. As LCs align along the electric field, the dye with a similar molecular shape is also led to align with LCs. Accordingly, the absorption of the GH can be controlled by an electric field. The extinction ratio depends on the order parameter of the host LC material and the dichroic parameter of the dye. Compared with TN and ECB modes, the GH mode has less wavelength dependency due to optical interference or retardation.

6.3.4 Projector application

For projector application, a high extinction ratio (contrast ratio) of more than 500 of amplitude modulation is required. Figure 6.15 depicts the optical system widely used in LC projectors. There are three amplitude modulation SLMs surrounding a dichroic prism [18].

Moreover, recent LC projectors require a high-density SLM, called a light valve. In the projection system, a lot of optical apparatus has been used, such as a colour combination and separation system, a polarization-converter system and a uniform illumination system (figure 6.16). The system can make projected images 1.5 times brighter because both s- and p-polarized light can be utilized.

For light valves, TFT or MOS transistor driven TN, ECB, PDLC, reflective TN-ECB, etc., can be used. They are listed in table 6.1.

6.4 Phase modulation and applications

Phase type SLMs with real time operation are required for various kinds of application. For example, there are active optical elements, beam steering, image recognition, optical processing, correlators and holographic imaging.

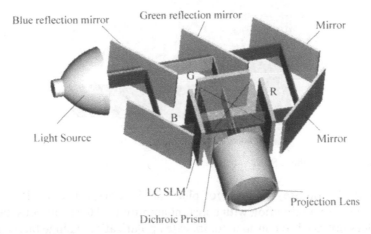

Figure 6.15. Optical system of an LC projector.

As a modulation medium, LC is a large birefringent material with high sensitivity for field 10^3 times larger than LiNbO$_3$. So far, LC SLMs that can offer a versatile optical phase modulation device of two dimensions are the only available device. Next, some typical devices and their applications are described.

6.4.1 Active matrix phase modulator

As mentioned above, active matrix addressing technologies for displays have been widely used and improve day by day. The technologies have high quality and versatility. However, there are some difficulties in using them

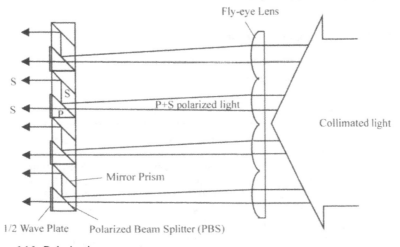

Figure 6.16. Polarization converter.

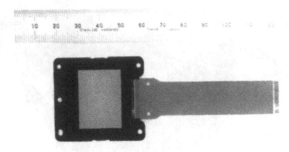

Figure 6.17. TFT-phase SLM.

in SLMs. If you use TN mode for phase modulators, it is difficult to obtain enough phase modulation with constant amplitude. This is obvious, because TN does not need to consider complicated polarization change but can use a linearly polarized component of output light. The simplest way to realize phase modulation is to use ECB mode that is aligned homogeneously in parallel [19].

6.4.1.1 TFT-driven spatial phase modulator

Figure 6.17 is an example of TFT-driven spatial phase modulators, whose specifications are shown in table 6.2. The device incorporates nematic LC due to its stability and reliability. The birefringence of the LC is $\Delta n = 0.166$. In a simple theoretical calculation, 2π modulation requires more than 3.8 μm LC thickness, and therefore we fabricated the panel with 4.5 μm thickness. Both substrates of the SLM in figure 6.17 are rubbed in parallel along the y direction. The homogeneous LC alignment in the y direction is obtained accordingly.

6.4.1.2 Phase modulation characteristics

Figure 6.18 shows phase modulation characteristics as a parameter of the wavelength. Video voltage indicates actual voltage applied to an individual

Table 6.2. Specifications of TFT-phase SLMs.

	1.3″ VGA	0.9″ VGA
Size	26.9 × 20.2 mm	18.5 × 13.9 mm
Pixel pitch	42 μm × 42 μm	23 μm × 23 μm
Number of pixels	640 × 480	640 × 480
Aperture ratio	64%	50%
δ (phase modulation)	2.5π	2π
Cell gap	4.5 μm	3.5 μm

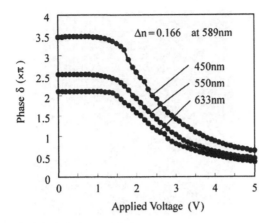

Figure 6.18. Phase modulation characteristics.

LC pixel through a TFT element. Because of the wavelength dependency of the LC birefringence, they inevitably show variations. A modulation depth of more than 2π is attained in cases of less than 550 nm. As can be seen in figure 6.18, there are wavelength dependences due to the dispersion of LC birefringence.

6.4.2 Holographic applications

6.4.2.1 Kinoform reconstruction

Using the SLMs, an experiment for active optical elements is now introduced: kinoform reconstruction. Kinoform is similar to hologram except for the term of amplitude, where kinoform retains a constant value. Hence, it is suitable to confirm the ability of the phase modulation characteristics. The kinoform patterns computed using the iterative method are shown in figure 6.19. Figure 6.20 shows images reconstructed by using an He-Ne laser and an 800 mm convex imaging lens. Speckle noise and minus-order image are well suppressed. Diffraction efficiency is assumed to be a value close to the square of the SLM's aperture ratio, because phase distribution can be displayed with high fidelity. Hence, diffraction efficiency of about 40% for the aperture ratio of 64% is expected. In addition, using the TFT driving method with analogue video signals, continuous phase modulation, in other words 'blazed grating', is easily obtained, and can provide high diffraction efficiency.

6.4.2.2 Interference and diffraction effects

There are some unexpected effects, interference and pixel diffraction [20]. The interference is caused by layers between substrates, and the diffraction

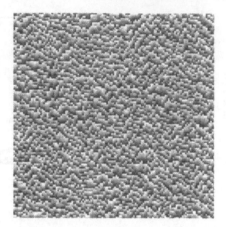

Figure 6.19. Kinoform pattern.

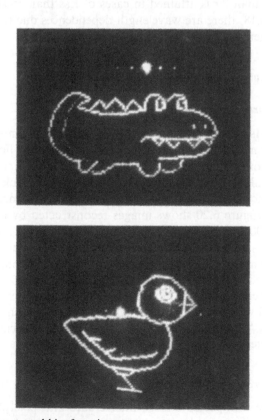

Figure 6.20. Reconstucted kinoform images.

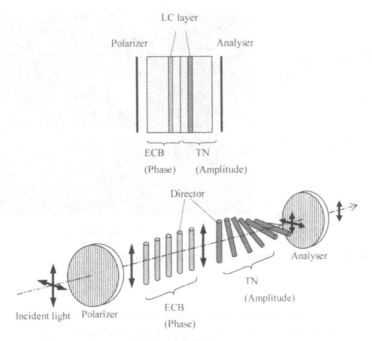

Figure 6.21. Spatial wavefront modulator (SWM).

is caused by a non-uniform electric field across each pixel, which is shown in figure 6.9. Potential distribution is disturbed in the vicinity of the gate electrode. LC responds dynamically with the non-uniform field. Consequently the effective refractive index in the pixel area shows gradient, and a blazed grating in the matrix's pixel pitch appears at ON state. Of course, the pixel regular structure causes the apparent diffraction, by which the higher order images, shown in figure 6.20, are formed. As explained above, some active matrix such as IPS has positively used the lateral field effect. Even in the case of the longitudinal field mode, the lateral field caused by fine pixel structures must be taken into account [21].

6.4.3 Spatial wavefront modulator (SWM)

The spatial wavefront modulator (SWM) shown in figure 6.21 was developed for a real-time computer-generated hologram (CGH) reconstruction at almost the same time as the above-mentioned phase modulators [22]. The SWM is composed of a TN-mode SLM and an ECB-mode SLM. The TN is for amplitude modulation and the ECB is for phase, respectively. Two SLMs are attached across both thin glass substrates, and each SLM is independently driven with each TFT active matrix. Accordingly, SWM can

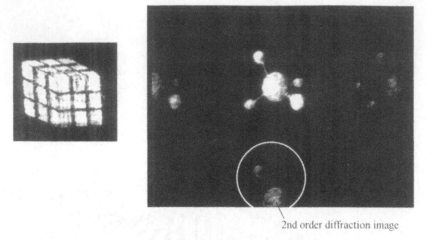

2nd order diffraction image

Figure 6.22. Three-dimensional reconstruction by using SWM.

complete complex amplitude modulation in real time, and it would be an ideal SLM that can generate any wavefront you want. Figure 6.22 is a result of the holographic reconstruction for the proof of its modulating performance.

Complex amplitude modulation is represented by

$$C = (A/B)\exp(i\alpha - i\beta),$$

where $A\exp(i\alpha)$ denotes the object wave and $B\exp(i\beta)$ the reference wave, respectively. Both waves are complex amplitude, including amplitude and phase terms.

When the reference wave is illuminated on C, the object wave can be reconstructed completely. If the reference wave is, for example, a plane wave, C is easily calculated. Next, C is divided into the amplitude and phase terms, which are transformed into the display data of the TN and ECB SLMs.

In the case of figure 6.22, the complex amplitude data are calculated as follows. Many sliced plane images in different depths are changed to CGH data by Fresnel transformation and accumulated. On the other hand, it is also practicable to treat the object as the assembly of point light sources, which is directly transformed to accumulated CGH data.

A motion CGH video by using the SWM was demonstrated. The animation amplitude and phase data are calculated in advance, and stored in an optical disk recorder. When fast hardware to handle 10^6 points fast Fourier transform (FFT) calculation within a frame time is available, a real-time vivid CGH display will be possible.

6.4.4 Optical addressed phase modulator

A phase-type SLM by optical addressing, described previously in figure 6.12, is offered by Hamamatsu Photonics [23] and has been used for an optical recognition system and optical modelling system as well as imaging.

6.5 Scattering and deflection modulation

Scattering could be regarded as random deflection of light. In this sense, scattering and deflection are described in this section.

In the historical view of LC displays, scattering effects appeared earlier than TN. The first was the dynamic scattering mode, which is based on dynamic disorder caused by ions' drift in an LC layer. Unfortunately it was used for just a few years due to poor reliability. Then the TN mode, based on field effect, was invented, and took the major place in display applications.

6.5.1 PDLC

After the initial developments, polymer dispersed in liquid crystal (PDLC), made an appearance. PDLC is a composite system with polymer and LC, and shows electrically controlled scattering, which is a kind of field effect like TN. The scattering principle of PDLC is based on induced mismatch of refractive index, which is caused by mismatch between the birefringence of LC and polymer index [24]. In principle, it requires no polarizer, and differs from TN or ECB that utilize polarization change in birefringence optical media. A polarizer is not suitable for reflective displays, which have the limitation of only using ambient light.

In addition, PDLC is a unique technology that can form artificial three-dimensional structures in an LC cell. Ordinary PDLC has LCs confined in a polymeric matrix. The device which has a polymer-linked structure was developed by us, and called the IRIS (internal reflection inverted scattering) mode [25]. Holographic PDLC has a volume holographic structure. These structures affect the light propagation through the composite system, so an applied field can modulate scattering and/or diffraction.

In common display applications, we usually use backscattering, which means that the light is mainly scattered in the opposite direction to incidence. In some cases, particularly in applications for SLMs or light valves, forward scattering has been also used.

First, the backscattering type is described. The first current PDLC technology appeared as nematic encapsulated aligned in polymer (NCAP) in 1981. Then polymer network liquid crystal (PNLC) and other related PDLC technologies have been developed. All of them have a rigid polymer

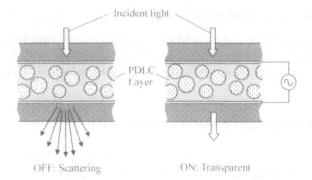

Figure 6.23. Ordinary PDLC.

morphology like a three-dimensional net or an Emmental cheese and LCs contained in them. Figure 6.23 shows ordinary PDLC structures. Due to LCs randomly oriented in the polymer structure, they exhibit strong back-scattering and normally scattered operation, namely scattering at the OFF (no field) state, where scattering does not depend on polarization of light. At the ON (applied field) state, LC aligns along the electric field, getting the index of LC close to that of polymers, and scattering disappears. Naturally, by modifying device parameters, the scattering characteristics can be changed as required.

As an example of the forward scattering type, I introduce IRIS, which is the first full-colour reflective PDLC display [25]. IRIS shows almost no absorption itself and more than 90% transmissivity in the OFF state and forward scattering in the ON state. The polymer structure is aligned along LC alignment during UV polymerization. The structure is shown in figure 6.24. Compared with the common PDLC, the polymer morphology

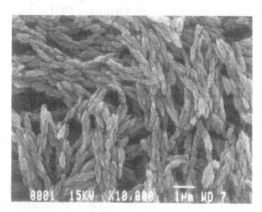

Figure 6.24. Polymer morphology of IRIS.

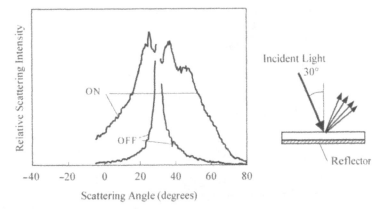

Figure 6.25. Scattering characteristics of IRIS.

of IRIS is in accordance with LC orientation in the whole PDLC system. This process is a unique and stimulating technology because LC alignment enables polymer morphology to be used to control the three-dimensional polymer structure. For instance, adding a small amount of optical chirality to the system, a twisted layer polymer structure can easily be obtained. An IRIS-PDLC system would also be useful for the study of phase separation. After polymerization, the system becomes transparent with well-matched refractive index to each other. This is the OFF state. When an electric field is applied, the liquid crystal becomes re-aligned along the electric field, and scattering causes refractive index mismatching between the polymer and the liquid crystal. Thus IRIS shows an inverted scattering behaviour. The system shows relatively sharp forward scattering characteristics and low driving voltage compared with ordinary PDLCs. Figure 6.25 shows typical angular scattering characteristics of IRIS. To utilize forward scattering for displays, we combined the system with a reflector, which can also control the scattered light direction. This configuration conveys higher brightness than common reflective TN and enables a RGB micro colour filter to be used for colour representation.

6.5.2.1 *Scattering and Schlieren optics*

If it is required to change scattering or diffraction to amplitude modulation, selective optics such as Schlieren optics must be used. This system is based on the selection of a part of the distribution of light. There are three elements, a light source, an LC modulator and a selecting element, which can control the light distribution of each other. This optical system has been mainly used for projection systems [26]. In some cases of scattering displays, the iris of the observer's eyes plays the role of a Schlieren aperture.

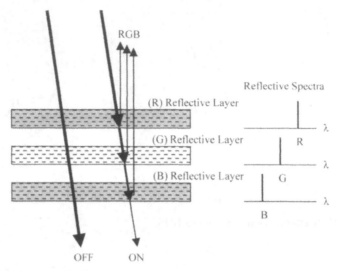

Figure 6.26. Holographic PDLC.

6.5.2 Holographic PDLC

In the photo-polymerization process, holographic PDLC (H-PDLC) uses a method of interference UV exposure. Then a hologram of polymer is formed in the composite layer. This is another kind of controlled three-dimensional structured PDLCs. As easily understood, strong directional reflection with a wavelength selective property can be achieved as in the case of volume holograms. In other words, the formation of the H-PDLC is equivalent to a photo-polymer used in hologram recording (figure 6.26). The features of H-PDLC are versatile for colour-reflective displays and directional light control [27].

References

[1] http://www.displaytech.com
[2] Alt P M and Pleshko P 1974 *IEEE Trans.* **ED21** 146
[3] Ruckmongathan T N 1988 *Proc. 8th International Display Research Conference* p 80
[4] Scheffer T J and Clifton B 1992 *SID '92 Digest* 228
[5] Shimoda T and Inoue S 1999 *IEDM Tech. Digest* 2289
[6] http://www.kopin.com
[7] Morozumi S, Ohta T, Araki R, Sonehara T, Kubota K, Nakazawa T and Ohara H 1983 *Proc. Japan Display* 404
[8] Oh-e M, Ohta M, Aratani S, and Kondo K 1995 *Proc. 15th International Display Research Conference* 577, 707
[9] Lee S H, Lee S L and Kim H Y 1998 *Proc. 18th International Display Research Conference* 371

[10] Melchior H, Kahn F J, Mayden D and Fraser D B 1972 *Appl. Phys. Lett.* **21**(8) 392

[11] Beard T D, Bleha W P and Wong S-Y 1973 *Appl. Phys. Lett* **22** 90

[12] Sterling R D, Te Kolste R D, Haggerly J M, Borah T C and Bleha W P 1990 *SID '90 Digest* 327

[13] Sonehara T and Kobayashi H 1990 *Proc. SPIE* **1254** 191

[14] Takizawa K, Kikuchi H, Fujikake H and Namikawa Y 1993 *Opt. Engng* **32** 1781

[15] Stanley M, Conway P B, Coomber S, Jones J C, Scattergood D C, Slinger C W, Bannister B W, Brown C V, Crossland W A and Travis A R L 2000 *Proc. SPIE* **3956** 13

[16] Haven D, Whitlow D and Jones T 1986 *SID '86 Digest* **372**

[17] Bawmann E 1953 *J. Soc. Motion Picture Television Engng* **60** 344

[18] Morozumi S, Sonehara T, Kamakura H, Ono T and Aruga S 1986 *SID '86 Digest* 375

[19] Sonehara T and Amako J 1997 *OSA TOPS* **14** 165

[20] Davis J A, Tsai P S, Cottrell D M, Sonehara T and Amako J 1999 *Opt. Engng* **38** 1051

[21] Onozawa T 1990 *Japan J. Appl. Phys.* **29** L1853

[22] Sonehara T, Miura H and Amako J 1992 *Proc. 12th International Display Research Conference Japan* p 315

[23] Igasaki Y, Li F, Yoshida N, Toyoda H, Inoue T, Mukohsaka N, Kobayashi Y and Hara T 1999 *Opt. Rev.* **6** 339

[24] Drzaic P S 1995 *Liquid Crystal Dispersions* (Singapore: World Scientific)

[25] Sonehara T, Yazaki M, Iisaka H, Tsuchiya Y, Sakata H, Amako J and Takeuchi T 1997 *SID '97 Digest* 1023

[26] Dewey A G 1977 *IEEE Trans.* **ED24** 918

[27] Tanaka K, Kato K, Date M and Sakai S 1995 *SID '95 Digest* 267

Index

Abbreviations

α-Si	amorphous silicon
α-Si:H	hydrogenated amorphous silicon
a-Si	amorphous silicon
AD	azo-dye
ADA	Anomalous Diffraction Approach
AFLC	AntiFerroelectric Liquid Crystal
AM	Active Matrix
AO	Adaptive Optics
B-OASLM	Bipolar-operational OASLM
BPOF	Binary Phase-Only Filter
BSO	$Bi_{12}SiO_{20}$ photoconductive crystal
CCD	Charge Coupled Device
CD	Compact Disc
CGH	Computer Generated Hologram
CRT	Cathode Ray Tube
CW	Continuous Wave
DBS	Direct Binary Search
DHF	Deformed Helix Ferroelectric (liquid crystal)
DLC	Discotic Liquid Crystal
DM	Deformable Mirror
DMF	n,n-dimethylformamide
DPR	Double Phase Retarder
DVD	Digital Versatile Disc (originally Digital Video Disc)
EA	Electrically Addressable
EASLM	Electrically Addressable Spatial Light Modulator
ECB	Electrically Controlled Birefringence
EDFA	Erbium Doped Fibre Amplifier
EO	Electro-Optical
FFS	Fringe Field Switching
FFT	Fast Fourier Transform
FLC	Ferroelectric Liquid Crystal

FP	Fabry–Pérot
FT	Fourier Transform
FWHM	Full Width Half Maximum
GDLC	Gel-glass Dispersed Liquid Crystal
GH	Guest–Host
GRIN	Graded Index
H-PDLC	Holographic Polymer Dispersed Liquid Crystal
HLPP	Hybrid LPP
HOE	Holographic Optical Element
HPDLC	Holographic Polymer Dispersed Liquid Crystal
HRPLC	Homeotropic Reverse-mode Polymer-LC
HTN	Hybrid Twisted Nematic
HWP	Half Wave Plate
I	Isotropic phase
IPS	In-Plane Switching
IR	Infrared
IRIS	Internal Reflection Inverted Scattering
ITO	Indium Tin Oxide (In_2O_3:Sn)
JTC	Joint Transform Correlator
LB	Langmuir–Blodgett
LC	Liquid Crystal
LCD	Liquid Crystal Display or Liquid Crystal Device
LCOS	Liquid Crystal On Silicon
LCP	Liquid Crystal Polymer (or polymeric liquid crystal)
LCPC	LC–Polymer Composite
LCTV	Liquid Crystal TV
LLC	Lyotropic Liquid Crystal
LMWPC	Low Molecular Weigh Photo-cross-linkable Composites
LPP	Linear Photo-polymerized Photo-alignment
MEMS	Micro-Electro-Mechanical System
MIM	Metal Insulator Metal
MLA	Multi Line Addressing
N	Nematic phase
n-CB	alkyl-cyano-biphenyls (e.g. 5-CB, 6-CB, etc.)
N–I	Nematic to Isotropic (phase transition)
N*	chiral Nematic
NA	Numerical Aperture
NCAP	Nematic Curvilinear Aligned Phase (or eNCAPsulated liquid crystal, or nematic encapsulated aligned in polymer)
NLC	Nematic Liquid Crystal
OA	Optically Addressable
OASLM	Optically Addressable Spatial Light Modulator
OFT	Optical Fourier Transform
OPD	Optical Phase Difference

P-PIPS	Photo-initiated PIPS
PBS	Polarizing Beam Splitter
PC	Personal Computer
PDCLC	Polymer Dispersed Chiral Liquid Crystal
PDCN	Polymer Dispersed Chiral Nematic
PDFLC	Polymer Dispersed Ferroelectric Liquid Crystal
PDL	Photonic Delay Line
PDLC	Polymer Dispersed Liquid Crystal
PDLC-LV	Polymer Dispersed Liquid Crystal Light Valve
PDLC/LM	Polymer Dispersed Liquid Crystal Light intensity Modulator
PDLC/PM	Polymer Dispersed Liquid Crystal light Phase Modulator
PDP	Plasma-discharge Display Panel
PEM	Photo-Elastic Modulator
PI	polyimide
PIBMA	poly-(isobutyl-methacrylate)
PIPS	Polymerization Induced Phase Separation
PMF	Polarization Maintaining Fibre
PMMA	poly-(methyl-methacrylate)
PMN	lead manganese niobate
PNLC	Polymer NLC
PoLiCryst	Polymer–Liquid Crystal
PSCT	Polymer Stabilized Cholesteric Texture
PSF	Point Spread Function
PSLC	Polymer Stabilized Liquid Crystal
PV	Peak-to-Valley
PVA	poly-vinyl-alcohol
PVCN	poly-vinyl-cinnamate
PVDF	poly-vinylidene-fluoride
PVF	poly-vinyl-formal
PVMC	poly-vinyl-4-methoxyl-cinnamate
PZT	piezoelectric
QWP	Quarter Wave Plate
RGA	Rayleigh–Gans Approximation
RGB	Red–Green-Blur
ROM	Read-Only Memory
SBWP	Spatial BandWidth Product
SIPS	Solvent Induced Phase Separation
SLM	Spatial Light Modulator
SmA	Smectic A phase
SmA^*	chiral Smectic A phase
SmC	Smectic C phase
SmC^*	chiral Smectic C phase
SmC_A^*	chiral Smectic C_A phase
SmC_A	Smectic C_A phase

SNR	Signal to Noise Ratio
SPLC	Sidechain Polymer Liquid Crystal
SRAM	Static Random Access Memory
SSFLC	Surface Stabilized Ferroelectric Liquid Crystal
STM	Scanning Tunneling Microscope
STN	Super Twisted Nematic
SWM	Spatial Wavefront Modulator
T-PIPS	Thermally-initiated PIPS
TFD	Thin Film Diode
TFEL	Thin Film Electro-Luminescent
TFT	Thin Film Transistor
TIPS	Temperature Induced Phase Separation
TIR	Total Internal Reflection
TN	Twisted Nematic
TNE	Transient Nematic Effect
UV	Ultraviolet
VAN	Vertically Aligned Nematic
VCSEL	Vertical Cavity Surface Emitting Laser
VHR	Voltage Holding Ratio
VLSI	Very Large Scale Integration
WDM	Wavelength Division Multiplexing

9 780367 454494